T0269149

CAMBRIDGE LIBRARY COLLECTION

Books of enduring scholarly value

Earth Sciences

In the nineteenth century, geology emerged as a distinct academic discipline. It pointed the way towards the theory of evolution, as scientists including Gideon Mantell, Adam Sedgwick, Charles Lyell and Roderick Murchison began to use the evidence of minerals, rock formations and fossils to demonstrate that the earth was older by millions of years than the conventional, Bible-based wisdom had supposed. They argued convincingly that the climate, flora and fauna of the distant past could be deduced from geological evidence. Volcanic activity, the formation of mountains, and the action of glaciers and rivers, tides and ocean currents also became better understood. This series includes landmark publications by pioneers of the modern earth sciences, who advanced the scientific understanding of our planet and the processes by which it is constantly re-shaped.

On the Geological Structure of the Alps, Apennines and Carpathians

Few men were better placed to produce an authoritative study of Continental geology than Roderick Impey Murchison (1792–1871), President of the Geological Society and the Royal Geographical Society of London. Having conducted extensive fieldwork alongside Adam Sedgwick, in 1847 Murchison set out on a study tour that would change the manner in which geology was understood and debated. Delivered before the Geological Society and published in their *Quarterly Journal* in 1849, this paper challenged received wisdom as to the age and formation of the most impressive of geological phenomena. Covering crystalline and palaeozoic rocks, the Trias, iron mines, nummulitic rocks and fish slates, this landmark study and its numerous diagrams that illustrate it not only explain a geological subject, but also reveal the nature of nineteenth-century scientific scholarship. The book also contains a short supplementary paper on the distribution of surface detritus in the Alps.

Cambridge University Press has long been a pioneer in the reissuing of out-of-print titles from its own backlist, producing digital reprints of books that are still sought after by scholars and students but could not be reprinted economically using traditional technology. The Cambridge Library Collection extends this activity to a wider range of books which are still of importance to researchers and professionals, either for the source material they contain, or as landmarks in the history of their academic discipline.

Drawing from the world-renowned collections in the Cambridge University Library, and guided by the advice of experts in each subject area, Cambridge University Press is using state-of-the-art scanning machines in its own Printing House to capture the content of each book selected for inclusion. The files are processed to give a consistently clear, crisp image, and the books finished to the high quality standard for which the Press is recognised around the world. The latest print-on-demand technology ensures that the books will remain available indefinitely, and that orders for single or multiple copies can quickly be supplied.

The Cambridge Library Collection will bring back to life books of enduring scholarly value (including out-of-copyright works originally issued by other publishers) across a wide range of disciplines in the humanities and social sciences and in science and technology.

On the Geological Structure of the Alps, Apennines and Carpathians

More Especially to Prove a Transition from Secondary to Tertiary Rocks, and the Development of Eocene Deposits in Southern Europe

RODERICK IMPEY MURCHISON

CAMBRIDGE UNIVERSITY PRESS

Cambridge, New York, Melbourne, Madrid, Cape Town,
Singapore, São Paolo, Delhi, Tokyo, Mexico City

Published in the United States of America by Cambridge University Press, New York

www.cambridge.org
Information on this title: www.cambridge.org/9781108072564

This edition first published 1849
This digitally printed version 2011

ISBN 978-1-108-07256-4 Paperback

ON THE

GEOLOGICAL STRUCTURE

OF THE

ALPS, APENNINES AND CARPATHIANS,

MORE ESPECIALLY TO PROVE A TRANSITION FROM SECONDARY TO TERTIARY ROCKS, AND THE DEVELOPMENT OF EOCENE DEPOSITS IN SOUTHERN EUROPE.

BY

SIR RODERICK IMPEY MURCHISON, G.C.S. F.R.S. G.S. L.S.,

Hon. Mem. R.S. Ed., R.I. Ac., Mem. Imp. Ac. Sc. St. Pet., Corr. Mem. Ac. France, Berlin, Turin, &c. &c.

LONDON:

PRINTED BY RICHARD AND JOHN E. TAYLOR,

RED LION COURT, FLEET STREET.

1849.

THE

QUARTERLY JOURNAL

OF

THE GEOLOGICAL SOCIETY OF LONDON.

PROCEEDINGS

OF

THE GEOLOGICAL SOCIETY.

DECEMBER 13, 1848.

Thomas Josiah Lang, Esq., and Charles Brumell, Esq., were elected Fellows of the Society.

The first part of the following communication* was read:—

1. *On the Geological Structure of the* ALPS, APENNINES *and* CARPATHIANS, *more especially to prove a transition from Secondary to Tertiary rocks, and the development of Eocene deposits in* SOUTHERN EUROPE. By Sir RODERICK IMPEY MURCHISON, F.R.S. G.S. L.S., Hon. Mem. R.S. Ed., R.I. Ac., Mem. Imp. Ac. Sc. St. Pet., Corr. Mem. Ac. France, Berlin, Turin, &c. &c.

INTRODUCTION.

THE numerous mineral distinctions of the various rocks composing the Alps, and their separation into more or less crystalline masses, were the chief objects of the researches of the illustrious De Saussure; and some time elapsed before it was thought possible to bring these mountains into anything like a comparison with the sedimentary deposits of other parts of Europe, the determination of which

* Part second was read on January 17, 1849.

had been established by their normal order of position and their imbedded organic remains. As soon however as Brochant (1808) declared his belief, that large crystalline masses of the Central and Savoy Alps, which had previously been considered of primary age, belonged to the earlier sedimentary or transition period, a new field of research was prepared; and Dr. Buckland made a still more important step, in a very able essay, wherein he boldly synchronized, in a general manner, the so-called transition rocks of Brochant, with our secondary British types*. Stimulated by such examples, and also by the researches of Brongniart, Von Buch, É. de Beaumont, Boué, Lill von Lilienstein and others, Professor Sedgwick and myself published our views in a memoir in the Transactions of the Geological Society†, accompanied by a general geological map of the Eastern Alps. Since that period, however, much progress having been made, by applying to this chain the more accurate knowledge of the order of equivalent formations, I had the strongest desire to revisit my old ground, to compare it with those regions of the Alps formerly unexplored by me, yet rendered classic by the discoveries of my contemporaries, and to correct any erroneous views I might have entertained. The great stimulus to my researches was, however, that I could not reconcile some of the phænomena I had formerly seen with the view of succession adopted in nearly every work and map of modern times, which represent the so-called cretaceous deposits of the Alps and Italy as being succeeded at once by the younger tertiary strata, almost to the entire exclusion of the eocene or older tertiary. One small tract only (the Vicentine) was supposed by some authors to be of lower tertiary age, whilst others even classed it with the chalk. I felt as certain as when we wrote our memoir, that however Professor Sedgwick and myself might have erred in regard to the age of the Gosau deposits, there were still good evidences of the transition from secondary to tertiary on which we had insisted, and which could not be put aside nor overlooked. For example, I was convinced, that there could be no mistake in the sections on the flank of the Venetian Alps near Bassano, which I presented to this Society before I explored the Austrian Alps,—sections that pointed out in the clearest manner the passage from the surface of the chalk into the *oldest tertiary* strata, and from them into newer deposits with subapennine

* See Annals of Philosophy, an. 1821, vol. xvii. p. 450. It is also but justice to the late Mr. Bakewell, to state that in examining the Alps of Savoy and the Tarentaise in the same summer as Dr. Buckland, he arrived at a similar conclusion (see Travels in the Tarentaise and various parts of Grecian and Pennine Alps, vol. ii. p. 410). In relation to my own researches I may now state, that in the year 1829 I went along the Maritime Alps, and afterwards, by Turin, to the Vicentine, with Sir C. Lyell. In the autumn of the same year I made the Bassano section and traversed the Tyrolese Alps. In 1829 Professor Sedgwick and myself examined the Eastern Alps, Styria and Illyria. In 1830 I returned alone to the Eastern Alps, and did not revisit them until 1847. In 1843 I made an excursion from Cracow to the Carpathian chain with Professor Zeuschner, and in the years 1847 and 1848 I was chiefly occupied in collecting data for this memoir.

† Vol. iii. Second Series, p. 301; and Phil. Mag. and Ann. of Phil. N. S. vol. viii. Aug. 1830.

shells*. These indisputable data were in fact the groundwork of the opinion afterwards applied to the Austrian Alps in natural sections, amidst some of the interior valleys, as well as upon their northern flank. Again, it was impossible to consider the shelly deposits of the Vicentine in any other light than older tertiary deposits, as laid down by Brongniart; and if they were of that age, they must, we argued, have equivalents in other parts of the Alps. In relation, even, to the deposits of Gosau, we then recognized, that their lower shelly beds were cretaceous by their fossils; but influenced both by the presence of an overwhelming quantity of associated gasteropoda, which usually abound in tertiary deposits (said to be of that age by conchologists), and also by the facies of the soft and incoherent deposits, which were so strikingly contrasted with the subcrystalline secondary rocks on which they reposed, we concluded that the upper shelly portion of the group also represented a transition from cretaceous to supracretaceous rocks, analogous to that seen on the flank of the Venetian Alps. My last visit to Gosau in 1847† has convinced me that my former view must be abandoned. I now believe that the marly and earthy fossiliferous beds of that valley are the equivalents of the gault, upper greensand and lower chalk‡. But if the shelly portion of the Gosau deposits proved to be cretaceous, the sections of Bassano and Asolo remained, as well as those of Untersberg and Kressenberg, to establish the existence of other and superior strata. And even when I say, that the Gosau deposits are essentially cretaceous by their fossils, I must guard against the inference that the overlying sandstones and schists of that valley are also of that age. The principal change of classification I have to make, is in respect to the comparison formerly suggested (though then not without considerable doubt), that the great band of green sandstones, impure limestones, and calcareous shale, &c. which occupy the external zone of the north-eastern Alps under the name of "Flysch" or Vienna sandstone, was the representative of the greensand and chalk of England and France. It is needless now to explain all the reasons for having embraced an opinion, which my colleague and self shared in common with other geologists of that day. In the absence of fossils, we could not, indeed, avoid being somewhat guided by mineral characters, particularly in the Eastern Alps, where the whole of this green sandstone zone abruptly succeeds to masses of what was then termed the "Alpine Limestone," the higher portion of which was considered by our precursors to represent the Upper Jura.

Once impressed with the conviction that the great greensand group succeeding to the supposed jurassic rocks was cretaceous, and finding nummulites associated with it, we naturally concluded that these

* See Phil. Mag. and Annals, with coloured sections, June 1829, and Proceed. Geol. Soc. Lond. vol. i. p. 137.

† On this occasion I was accompanied by M. de Verneuil.

‡ M. Boué argued that the fossil beds of Gosau were of the age of the lower greensand, whilst my colleague and self considered that these beds were both cretaceous and lower tertiary. M. Boué, as well as myself, now considers the nummulitic and flysch rocks as supracretaceous.

fossils were connected with both the older tertiary and the younger secondary deposits; an opinion from which I entirely recede. I am now persuaded that no form of the genus *Nummulina* (D'Orb.) occurs in the Alps in beds below the surface of the chalk, or its equivalent. Geologists must recollect, however, that at the period when we wrote, the development of the lower portion of the cretaceous system in Southern Europe had not even been commenced. The Neocomian formation was unknown, and no one dreamt that the thick outer coat of the subcrystalline alpine limestone, then considered to be of the age of the upper oolite, would prove to be the equivalent of the slightly coherent beds of sand and shale known as English "*Lower Greensand*!" The few secondary fossils we could then detect, in any rocks above strata containing liassic and jurassic species, were typical of the cretaceous epoch, and thus, putting aside dislocations, we supposed that the anomalous group called "flysch," containing so much greensand, and which as a whole was interpolated between the supposed jurassic rocks and the known tertiary deposits, might be of the same age as other sandy marls and calcareous bands, often also charged with green-earth, in which we found cretaceous fossils. Portions of the deposits of Gosau, as well as those on the northern flank of the Untersberg, had also to a great extent the characters of "flysch," and hence we supposed that such patches as contained cretaceous fossils were simply "oases" in a great secondary greensand succession.

I have now satisfied myself, that the great mass of the so-called flysch is the superior portion of the nummulitic "Terrain," and that the lowest beds with nummulites are completely above all those rocks which are the equivalents of the white chalk of northern Europe. In demonstrating this by absolute sections, I will further show, that between the representative of the chalk and the lowest nummulite limestone, there are beds, sometimes of considerable dimensions, which, whether marls, green sandstones, or impure limestone, exhibit that true transition I formerly insisted on as occurring between the secondary and tertiary rocks of the Alps.

The application of this classification to the Alps, Apennines and Carpathians, in all of which similar nummulitic limestones and sandstones occur, is loudly called for, seeing the discordant opinions which prevail respecting such deposits. In the valuable general map of Von Dechen, for example, the zone which is occupied by the flysch in the Eastern Alps is placed as the equivalent of the lower cretaceous rocks, without any representative of the chalk; and in defining the secondary boundary through Switzerland, the cretaceous system is omitted, the molasse being represented as in contact with the jurassic rocks. Yet this is the very region in which a most copious development of the *whole cretaceous system occurs*, overlaid by vast thicknesses of nummulite limestone and flysch. In the same map the deposits of the Vicentine are classed as lower tertiary, whilst they are, in truth, a peculiarly shelly portion only, of the same vast series of the supracretaceous rocks which embraces the nummulite limestone and flysch.

In Austria a new map of that empire has been published, in which the flysch of the very zone in question, or a large portion of it, is represented as Keuper. On the Italian face of the Alps and in the Apennines, some deposits, that I believe to be the same, are coloured as cretaceous, and are grouped (in the new map of Collegno) with all the deposits down to the lower greensand or Neocomian inclusive. This has been, indeed, the systematic view of most of the continental geologists. It has been chiefly adopted in pursuance of the opinions of M. Élie de Beaumont and M. Dufrénoy, who have coloured their admirable map of France on this principle. The conclusion of these authors is based upon the fact, that the nummulitic group, including the flysch of the Alps, has undergone all the movements which have affected the subjacent cretaceous rocks. Fully admitting that such are the physical relations, I nevertheless contend, that we cannot establish a comparative geological chronology between the strata of the north of Europe and those of the south, if after the evidences about to be submitted, we do not admit, that the group in question is truly lower tertiary, inasmuch as it lies above all rocks containing cretaceous or secondary fossils, is charged with an eocene fauna, and is succeeded in ascending order by formations filled with younger tertiary shells.

In the first portion of this memoir I give a general description, in ascending order, of the sedimentary rocks which constitute the whole chain of the Alps. After describing in succession the palæozoic and secondary formations, I point out the leading changes they have undergone in their range from the eastern to the western portion of the chain. The relations of the cretaceous and nummulitic rocks will then be discussed at greater length, followed by some data on the age and relations of the younger tertiary deposits of Switzerland; and this part will be concluded by descriptions of some of the principal fractures, inversions and contortions which these sedimentary strata have undergone.

A short sketch will give my present views of the succession on the north flank of the Carpathians, and explain the anomalies of the so-called Carpathian sandstone.

The third part, referring chiefly to Italy and the Apennines, will be terminated by a review of the organic remains and the order of the strata which establish the true age of the nummulitic group, not only in the south of Europe, but also in Egypt, Asia, and those vast regions of the globe over which it extends. A general *resumé* concludes the memoir.

Part I.

General Structure of the Alps.

It is now eighteen years since Prof. Sedgwick and myself pointed out that the chain of the Eastern Alps, when considered only in a general point of view, was of simple structure, in exhibiting a symmetrical succession of deposits from a crystalline centre through transition rocks now termed "palæozoic," flanked by grand secondary zones chiefly calcareous, which were externally environed by masses representing certain members of the tertiary deposits*. We stated, however, that when the geologist grappled with the detailed features of this chain, this apparent simplicity usually vanished, chiefly owing to the great movements of elevation and dislocation which it had undergone, and which frequently caused the younger formations to dip, or appear to dip, under those of more ancient date. But notwithstanding these difficulties, we then separated these Alpine rocks into a series of natural groups, admitting of at least a general comparison with the principal geological groups of England and other countries. Now, nearly all the general classification, as given in our Map, is still correct, and may stand at the present day. The examples, however, selected as proofs of the cretaceous and supracretaceous relations are inaccurate in the north-eastern Alps; and hence, though the legend or order of colours is on the whole right, its application to parts of the map must be changed, together with some essential portions of our reasoning.

Crystalline and Palæozoic Rocks of the Central Axis.—It was in this great group (to which the term palæozoic has since been applied) that we noted the presence of abundance of Encrinites in the talcose and chloritic limestones of the Tauern Alp. We also specially adverted to the presence of species of British carboniferous Producti in the old rocks near Bleiberg in Carinthia †. In short, we showed the existence of those fossils in strata which on the one hand were connected with masses in a crystalline state, and on the other with younger fossiliferous formations. The glimpse which we then obtained of this phænomenon has been matured into certain induction by additional recent discoveries in respect to other and older palæozoic strata ‡. Fossils have been recently discovered by M. Erlach at Dienten near Werfen in a portion of these transition rocks, which M. v. Hauer has noticed. On these fossils being shown by that gentleman to M. de Verneuil and myself when we visited Vienna in the summer of 1847, we identified one of the forms with the *Orthoceras gregarium* and another with *Cardiola interrupta*, both well-known British Upper Silurian fossils, associated with the *Cardium gracile* (Münster), a shell which occurs at Feugerolles in the Silurian rocks of Normandy.

The limestones of the environs of Grätz, near the eastern extremity of the chain, contain fossils of Silurian or Devonian age; perhaps of both formations. Having inspected, in company with M. de Verneuil, a portion of this ground immediately adjacent to the city of Grätz, particularly as seen in the adjacent hill of Plautsch, it appeared to us that the mountain, having a chloritic sandy limestone for its base, passing upwards through sandstone and grits and limestones of dark grey and reddish colours, with separating courses of chlo-

* Trans. Geol. Soc. Lond. vol. iii. New Series, p. 301, and Phil. Mag. and Annals of Phil. vol. viii. Aug. 1830.

† See Phil. Mag. and Annals of Phil. vol. viii. Aug. 1830.

‡ The exact position of which is indicated in the new geological map of M. A. Morlot, entitled 'Geologische Uebersichtskarte der Oesterreichischen Alpen.'

ritic schist, is capped on the summit by dark grey, white-veined, fossiliferous limestone, on the surfaces of which many corals weather out. These corals are, *Gorgonia infundibuliformis*, *Stromatopora concentrica*, *Cyathophyllum explanatum*, *C. turbinatum*, *C. hexagonum*, *C. cæspitosum*, *Astræa porosa* (Goldf.), *Heliopora interstincta* (Bronn), *Favosites polymorpha* (var. *ramosa* of the Devonian rocks), *F. spongites*? &c. As most of these polypifers range from the Upper Silurian into the Devonian, it might be difficult to class the limestones of Plautsch by reference to them only. The rock also contains, however, *Pecten grandævus* (Goldf.), *Cyathocrinites pinnatus* (Goldf.), *Inoceramus inversus* (Münster), *Orthoceras regularis*, and *Goniatites*. We also detected a very striking large bivalve, which is not only seen in the slabs of the pavement of Grätz, but which we also found on the summit of the Plautsch, and which we had at first sight believed to be a Strigocephalus. A better specimen, however, led us to think it might prove to be a Pentamerus not remote from the *P. Knightii*. Until, therefore, more clear specific forms be found and examined, it is not possible at once to say whether the palæozoic limestone of Grätz be Lower Devonian or Upper Silurian. In extending researches from that immediate district to the surrounding country, in which M. Rosthorn has, I am told, already made discoveries which will soon be communicated to the public, Silurian fossils like those of the tract south of Werfen may also be detected; and as the presence of carboniferous Producti has long been known near Bleiberg in the Carinthian Alps*, we shall then have the exhibition at intervals along both sides of the chief watershed of the Eastern Alps of sufficient reliquiæ of the palæozoic deposits to convince us of the former existence of considerable masses of sediment of that age. In the meantime, we have ample data to affirm, that large portions of the tract, coloured purple to indicate transition rocks on the map of my coadjutor and self, are occupied by rocks of true palæozoic age, which in many parts have passed into a crystalline state.

When, however, the geologist follows these palæozoic rocks upon their strike to the W.S.W., he perceives that the action of metamorphism has been much more developed in them. Already to the west of the Gastein Alps and the Grosse Glockner, the masses lying between the granitic or gneissose centre and the flanking walls of secondary limestone are found to be chlorite, talc and mica schists, in none of which have any traces of fossils been yet detected. In travelling in 1847 through portions of the mineral axis both to the east and west of the meridian of Innspruck, in company with M. Leopold von Buch and M. E. de Verneuil, I was forcibly struck with their great change in mineral succession, as compared with the more eastern range of the same masses; for chlorite and mica schists, assuming in parts almost the characters of gneiss, range up to the secondary limestones with scarcely any place for intermediate strata.

* A collection of the Bleiberg fossils having been shown to M. de Verneuil and myself at Vienna in 1847, we recognized therein not less than eight or ten species as belonging to the true carboniferous system of the palæozoic deposits.

These crystalline schists in the gorge of the Fünster-Münster Pass are seen to be permeated by thin courses of brilliantly white dolomite*, seeming to indicate, that in the great metamorphism to which these rocks had been subjected, thin courses and veins of limestone which were subordinate to them, had been transformed into a network of dolomite.

In following the watershed of the Alps from Austria into Switzerland and thence into the Savoy Alps, it becomes apparent that the zone of metamorphism widens. Not only is the place of those rocks, which in their eastern prolongation are palæozoic, taken by crystalline masses, but the metamorphism† has so extended, if I may thus speak, laterally from centre to flanks, as to affect in numberless instances the middle and even the younger secondary deposits, and in one or more tracts, as will be hereafter shown, has even converted into a crystalline state the strata called flysch, which I now consider to be of tertiary age.

No vestige of any fossil palæozoic animal has yet been brought to light in the Western Alps,—a fact, indeed, which accords with the phænomenon on which I am insisting, viz. that on the west the Alps have undergone a more intense degree of metamorphism than on the east. I shall have occasion to return to the consideration of this subject in speaking of those rocks of Savoy which contain belemnites and coal-plants.

In thus briefly touching upon the palæozoic rocks, in order to show that they have a distinct and indisputable existence only in the Eastern Alps, I ought to add that even there no traces have yet been discovered of the uppermost portion of such palæozoic series. The Permian system, so copiously developed in Northern Europe and especially in Russia, seems, in fact, never to have been deposited in Southern Europe.

The Trias.—The group, which crops out at various points from beneath the great mass of secondary limestones of the Eastern Alps, and is interposed between them and the palæozoic rocks above-mentioned, was correctly placed by Professor Sedgwick and myself in the parallel of "Keuper, Muschelkalk and Bunter Sandstein ‡." For this,

* These veins, though so white on fracture, weather yellow under the atmosphere and the action of water.

† I have no intention of going into any details respecting the extension of metamorphism from centre to flank in the Swiss and Savoy Alps. I had the great advantage of making an excursion last summer with M. Studer, who pointed out to M. P. Merian, M. Favre, and myself, the lateral extension of this phænomenon in the mountains which encase the glacier of Grindelwald. They have been called "coins" or corners of gneiss, that wedge into and invade the jurassic limestone which they overlap at great altitudes. The appearances conveyed no idea of wedges of any *pre-existing* crystalline rock of higher antiquity than the limestone, having been forced into the latter; but on the contrary were plain proofs to my mind, that an action of metamorphism ramifying laterally had invaded and altered the jurassic strata *in situ*.—See M. Studer's communication, Bull. Géol. Fr. vol. iv. p. 209.

‡ See the foreign synonyms of the legend attached to our Map. M. Morlot is in error in attaching to this red zone the name of "Rothliegendes?"—for that rock, which is a part of the Permian system, does not, as already stated, exist in the Alps.

however, we had no other grounds than that the mass was inferior to lias, and that besides salt and gypsum, it also contained a bivalve resembling forms known to us in strata of that age. Recently, however, this point has been satisfactorily established in respect to the Alps of the Tyrol, and I will here offer a few evidences of the age of the formation which fell under my notice in the autumn of 1847, when I accompanied MM. von Buch and de Verneuil to St. Cassian and the adjacent tracts, and also when we explored the same series around Recoaro, north of Vicenza.

The trias of the South Tyrol with which I am acquainted, consists of a group of sandstones, marls and limestones, the latter rarely in the state of dolomite, which ranges from E.N.E. to W.S.W., between the transition and crystalline rocks of the central axis (Brunecken, Brixen, &c.) on the north, and the great masses of alpine limestone (liassic and jurassic), which to the south, for the most part in the state of dolomite, range from the Ampezzo Pass to Botzen. This trias is peculiarly well exhibited in the Grödner valley to the east of the great road between Botzen and Klausen. The portion of this tract which lies around the little mountain village of St. Cassian*, at the limit of the German and Italian Tyrol, is that which has afforded the great variety of fossils first made known to naturalists by Count Münster, and since described by M. Klipstein. A great number of these forms being of new and unknown species, considerable doubt hung over the precise age of the deposit. This obscurity has been principally cleared away by an excellent short memoir of M. Emmerich, who, working out the details of a district rendered classical twenty-five years ago by the researches of Leopold von Buch, has clearly exposed the order of the strata, thus leaving little or no doubt, that the chief and peculiar group of fossils of those Alps belongs to the trias. Still the subject required confirmation, and M. von Buch being as desirous as myself of re-examining the tract, I had the good fortune to accompany him and M. de Verneuil thither. Ascending from the Eisach Thal at Atzwang, we passed under the grand dolomitic peaks of the Schlerns mountain, by Seiss and its bosses of melaphyre, to Castel Ruth. The plateau which there constitutes the base of all the overlying masses of limestones and marls, is a spotted red and green or true bunter sandstein, a very good building-stone, with strong courses of subordinate white sandstone. At St. Michael we examined the collection of fossils made by M. Clara, the venerable clergyman of that hamlet, in the strata which form slopes beneath the lofty escarpments of Paflatsch Berg, a promontory of the Seisser-Alp, to the south of his residence. Among these fossils we at once recognised the well-known *Trigonellites pes-anseris* of the muschelkalk (Myaphora or Trigonia) with many fragments of the stems of the lily encrinite, together with certain forms of Avicula and Posidonia†, and we were therefore at

* St. Cassian is upwards of 5000 English feet above the sea and near the head of the transverse valley, whose waters flowing from south to north fall into the Rienz west of Brunecken.

† The most remarkable of the Posidoniæ has been named by M. Emmerich *P. Claræ*, after the venerable pastor, who had discovered the *Trigonellites pes-*

once convinced that they had been derived from a band of true muschelkalk. As the fossils were procured from the band of limestone and shale which surmounts the red sandstone of the plateau upon which we stood, we had therefore already before us two members of the trias. In these valleys the muschelkalk is, however, not only based on red sandstone, but is associated with, and surmounted by red marls representing the Keuper, as seen in the face of the Mittag's Kogel and along various parts of the Schlerns and the Seisser-Alp, and near St. Ulrich in the Grödner-thal, where the whole of the trias is further overlaid by other limestones chiefly jurassic.

In proceeding from the Grödner-thal to Colfosco, by St. Christina through the pass of Sta Maria, the geologist cannot avoid being struck with the grand pyramidal and towering peaks of fantastic shape which the dolomite there assumes in the Lang-Kogel and other mountains*. The great vertical fissures and joints which traverse that rock must not, however, be confounded with the lines of true bedding, which are often more or less horizontal and undulating only, and which, though with difficulty observed by an unpractised geological eye, were visibly delineated before us by wreaths of snow which fell during an autumnal storm on the peaks and escarpments around St. Cassian. The pass of Colfosco, which shows clear sections of muschelkalk on the west, is further remarkable in affording fine examples of buttresses of black porphyry (melaphyre), which in one situation, west of Colfosco, is observable in absolute contact with highly dislocated, amorphous, pure white dolomite. I presume that some of this dolomite is of the age of the muschelkalk, because it is associated with certain beds of triassic grit called by M. Emmerich "Halobian sandstein." In descending the valley from St. Cassian by Stern and the Abtey-thal or to the north†, the trias

anseris since M. Emmerich's visit. The *Ammonites Johannis Austriæ* (Von Hauer) from the lower limestones of Halstadt has also proved to be one of the fossils of this deposit in the Tyrol.

* The peaks on the south side of this pass are termed Pissada Spitz, Masons Spitz, &c.

† In his work entitled 'Uebersicht über die geognostischen Verhältnisse Südtyrols, 1846,' Dr. H. Emmerich distinguishes the following Neptunian deposits in ascending order in this region:—1. Red sandstone. 2. Posidonia limestone, which he considers the same stratum that contains the *Trigonia vulgaris*, *Terebratula trigonella*, *Gervillia socialis* and *Encrinites liliformis* at Recoaro, and is the true muschelkalk. 3. Hornstein-führender kalk, a small and local deposit observed by M. Fuchs. 4. Halobian strata (black sandstone and calcareous schist). This rock (which has been termed grauwacke) is the lias of Klipstein, the Wengen deposit of Wissman, and the dolerite sandstone of Fuchs. 5. St. Cassian beds usually united with the Halobian sandstone. 6. Upper limestone with corals and brachiopods, in which, according to M. Fuchs, the fossils of St. Cassian occur at Sotto di Sasso, about 8000 English feet above the sea. Above all these come the jurassic dolomites. As to the *Halobian sandstone*, so called from the occurrence of the genus *Halobia*, it appeared to me to be a rock formed either during submarine volcanic action or out of the detritus of a plutonic mass. It is evidently a local deposit, of which there are no traces near Recoaro and Schio, where it is doubtless represented by other sandstones. The intermixture of fossils under the head of "St. Cassian" is accounted for by the fact, that they are collected in the rivulets which descend through jurassic as well as triassic deposits.

group is first clearly exposed with the dark grey or blackish "Halobian" sandstone, rising out from beneath the dolomitic rocks of the Heilige-Kreutz, and then is exhibited in several flexures along the sides of the gorge which leads to St. Martin. In this gorge, particularly near Poderova, vertical walls of muschelkalk throw off red marls occasionally gypseous, and pebbly conglomerates with many curvatures and flexures; the red ground being usually interlaced and associated with limestones. Lastly, the system is flanked to the north of St. Martin by crystalline schists (here very micaceous) which occupy the place of the transition and metamorphic rocks of the Central Tyrol, and range to the left bank of the river Rienz; for, on the right bank of this stream near Sonnenburg is seen the western end of one of the finest of Leopold von Buch's so-called ellipsoids of granite, the eastern extremity of which constitutes the striking Iffiger Spitz near Meran. The protrusion and juxtaposition of this grand mass of granite accounts, indeed, for the highly crystalline condition of the schists between it and the trias.

Trias of Recoaro and the adjacent tracts.—Between the zone of trias which occupies the valleys of Grödner, St. Cassian, &c., and the southern edge of the Tyrolese Alps, there are other ellipsoids of crystalline rock, which in their elevation have exposed considerable thicknesses of sedimentary deposits around them. The chief of these is the Cima d'Asti with its central granite and its accompanying crystalline formations in the Val Sugana, which I traversed rapidly on a former occasion. Further to the S.S.W. are other elliptical masses, as seen in the valleys of Leogre, Posina and Recoaro, which exhibit fundamental rocks of mica schist covered by an ascending series, in which the trias is the prominent formation. By the valley of Recoaro, I do not simply mean the gorge in which the baths and village of that name are situated, but also all the undulating district embracing Rovegliano and Communda, as well as both sides of the "Valle de Signori," on which the road from Schio and Rovereto proceeds by the pass of Corneto*. He who has but little time at his disposal may rest satisfied with the very clear exposition of the triassic strata which are exposed on the side of the Spitz mountain, immediately south of the village of Recoaro; where ascending from a base of mica or stea-schist traversed by trap dykes, some of which run almost horizontally with the strata, the observer will recognize the following ascending order, as exposed in the annexed woodcut (fig. 1).

Some doubt may exist as to the age of the bottom stratum (2) resting on the mica schist (1). It is a red and white spotted, micaceous sandstone with patches of coaly matter and carbonized vegetables, and some geologists may be disposed to consider it carboniferous. But it

* Having walked over this district by the mountain tract from the Corneto Pass to Recoaro, I afterwards revisited it, as well as the adjacent districts of the Venetian Alps, in company with the chief members of the Geological Section of the Venetian meeting of the "Scienziati Italiani," including not only my previous companions Von Buch and De Verneuil, but also the Marchese Pareto, M. de Zigno, Major Charters, M. Parolini, and Mr. Pentland. M. Trettenero explained the details at Recoaro, and M. Pasini became our leader in the region around Schio and in the Setti Communi, of which hereafter.

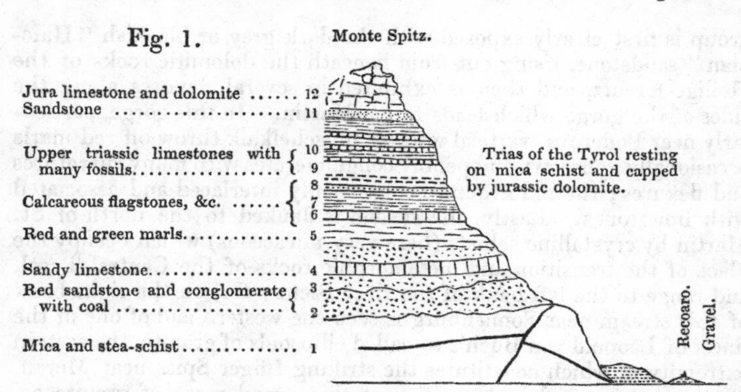

contains no plants on whose forms reliance could be placed, and it may prove to be a lower portion of the "bunter sandstein." Indeed, the whole stratum is of too feeble dimensions (not exceeding 40 to 50 feet) to require more illustration. It graduates upwards through calciferous sandstones (4) into red and green marls (5), which unquestionably belong to the trias, as proved by fossils found in them. The overlying beds of trias are various bands of limestone (6, 7, 8), one of which is slightly oolitic, which alternate with red marly and sandy beds; and whilst certain bivalves, such as Myacites, &c., occur in the lower flagstone strata, the upper masses (9, 10) have afforded the greatest number of good and peculiar muschelkalk fossils. It was in this band that Professor Brünner, jun., in a previous year, detected the beautiful Encrinite since named *E. gracilis* by V. Buch. From this band we also collected many fragments of the *E. liliformis* with *Terebratula vulgaris*, &c., as well as forms common to the lower strata of the group. Above this are sandstones (11) and jurassic dolomitic limestone (12).

In the deep ravines which lie to the north and west of Recoaro, similar successions of red sandstone, limestone and marls are observed. Fossils occur most abundantly in the ridges of shelly limestone east of Rovegliano (Communda Pass) and at Civelina, where slabs of flaggy limestone are absolutely covered with Trigoniæ (Trigonellites), with *Terebratula vulgaris, Myacites,* and other characteristic fossils of the muschelkalk; the deep denudation frequently exposing the same descending order. In one of the lower fossil beds of yellowish earthy limestone associated with reddish layers, our clever guide detected the *Spirifer fragilis* *.

With these very clear proofs of the full development of the trias in the localities already cited, there is every reason to believe, that the equivalents of the system (not, however, always fossiliferous) may be traced continuously along certain zones throughout the Eastern Alps, and particularly where red rocks and limestone rise out from beneath the great masses of alpine (lias and Jura) limestone, and repose upon

* This guide, Castellan, is recommended to geologists as an expert finder of fossils.

the transition or palæozoic rocks. In the general map before alluded to, I roughly sketched such a zone both on the north and south flanks of the transition rocks, and its presence has since been laid down in greater detail by M. Morlot in his useful general map of the Austrian Alps*; though I cannot admit his suggestion that such red rocks can be equivalents of the Rothe-todte-liegende.

The red sandstones (occasionally with certain yellowish limestones and also with salt and gypsum), which my colleague and self described in precisely the same geological position in the northern portion of these Alps, are, I have no doubt, of similar age to those described in the South Tyrol and Venetian Alps.

The researches of the palæontologist in the associated limestones have, indeed, to a great extent set that question at rest. M. von Hauer, jun., of Vienna, has shown that some of the fossils in the Salzburg Alps are identical with those which occur in the South Tyrol in the environs of Castel Ruth, St. Ulrich and St. Cassian; thus establishing the existence of true muschelkalk types in the northern zone, where they had not before been recognized. Among the fossils common to both tracts is the *Ammonites Johannis Austriæ* (Hauer).

Whether any true triassic plants occur in strata of that age in the *central* escarpments of the north-eastern Alps, is unknown to me. But the discovery of them in certain places near Waidhofen and Steyer, either in the middle of the area occupied by the secondary limestones or at their northern edge, has led to a classification on the part of the eminent mineralogist Haidinger, which, with every respect for him, I would suggest is not founded on a sound geological basis. The *Calamites arenaceus, Pterophyllum longifolium,* &c. (identified as such by Dr. Unger of Grätz) have, indeed, been found in a sandstone which dips under the liassic and jurassic limestones of the chain, and such keuper plants are therefore incontestably in their right position. Now, as this sandstone resembles some of the sandstones with impure limestones, that constitute a great zone geographically external to the whole of the alpine limestone, and to which the name of "flysch" has been applied, and of which the "Wiener sandstein" is the prolongation, M. Haidinger has identified the one with the other, and in consequence has recently coloured as "keuper," the whole of the flysch†. This zone of rocks is that which, often affected by great longitudinal faults, appears to dip under the alpine limestone, and had been classed as the uppermost secondary mass of the Alps. It is this very zone to which I am about presently to call attention in detail as belonging to much younger deposits. If the enormous thick accumulations of grits, limestones and fucoid schists to which I now allude, had a real existence beneath the lias where the strata are not inverted, then surely such rocks would somewhere be seen in the well-exposed natural escarpments near the base of those secondary rocks which lie in their normal positions. Such, indeed, is the position of the trias above described.

* See also the octavo volume accompanying the map by the same author, "Erlauterungen zur geologischen Uebersichtskarte der Nordöstlichen Alpen, von A. v. Morlot."

† See the new Geological Map of the Austrian Empire.

The view of M. Haidinger is, I apprehend, to be accounted for by an attempt to identify rocks by mineral resemblances, and by comparing strata in broken and inverted positions with others (supposed to be the same) which lie in their normal state. Finding certain sandstones which resemble the "Wiener sandstein" really plunging under the alpine limestone and lias, and containing keuper plants, and further seeing that the great mass of the Vienna grits external to the chain which are of similar appearance also *seem* to plunge under the same limestone ridge, a conclusion has been arrived at which I apprehend my distinguished friend M. Haidinger will abandon, as applied to most of the large and widely spread masses in question. For, if all the Vienna grits so intercalated between the great masses of secondary and tertiary rocks, be the representatives of the keuper, then all the "flysch" of Switzerland, the grits and sandstones on the flanks of the Carpathians, and the upper macigno of the Italians, may on such reasoning be similarly grouped. Now, although a band of true keuper sandstone with plants may crop out in the localities cited, it is *physically impossible* that all the great external zone to which I have alluded, and which, as I shall presently show, truly forms the last member in ascending order of the great chain of the Alps, can be referred to the keuper, a natural system which is now well understood in the Eastern Alps, and not one of whose fossils has ever been found in that external zone of Vienna sandstone, which forms the continuation of the Swiss and Bavarian "flysch" on all preceding maps*.

In the extension of the trias westward through the North Tyrol, its presence is only as yet recognizable to a very limited and doubtful extent. In short, it may be stated, that no discoveries have been made, either in the central Swiss Alps or in Savoy, which can lead us to think that the trias has had any existence there. Still, judging from the analogy of the Eastern Alps, it *is possible* that a spot or two may be found where the limestone is so little penetrated by metamorphic action as to have left some intelligible evidence of the triassic group. Leopold von Buch inclines, I believe, to the opinion that such a group will be detected.

* M. Morlot's map of the North-Eastern Alps is, as I conceive, quite correct in representing these Vienna grits to be a prolongation of the "flysch" of Switzerland, and in placing them in their true overlying position. But since its publication, M. Morlot has abandoned that opinion, and has adopted the view of M. Haidinger. See a brief notice (Sulla Conformazione geologica dell' Istria, Giornale dell' Istria, Nos. 61, 62, 1847). Again, in a communication to M. Haidinger on the position occupied by the "Wiener sandstein" or fucoid grits, he endeavours to show, that in following it from Istria into the interior of the Alps up the valley of the Isonzo, that formation is seen near Raibl to take, as he says, "its place between the lower muschelkalk (?) and upper alpine limestone," and thus represents the Keuper (Reports of the Meetings of the Friends of Science, Vienna, Haidinger, vol. iii. Oct. 1847, p. 334). Whilst these sheets are passing through the press, the author has received M. Morlot's detailed memoir and map of Istria, "Naturwissenschaftliche Abhandlungen, vol. ii. p. 257." His descending order is,—1. Eocene or nummulitic rocks; 2. Chalk; 3 Tassello. I have some difficulty in reconciling this order either with some of the sections or with the colouring of the map of Istria by M. Morlot.

Lower and Upper Alpine Limestones.

Liasso-Jurassic and Oxfordian.—Following the classification of M. Lill von Lilienbach, Professor Sedgwick and myself considered the alpine limestone of the Salzburg Alps as divided into two great masses, separated by shale, sandstone and limestone with certain saliferous deposits. In differing from some of our contemporaries in assigning a portion of the lowest of these fossiliferous limestones at Admet, to the age of the lias, I still believe we were correct. But, as before said, in the accurate determination of all the members of these lower secondary rocks, a portion of which is now ascertained to be triassic, so much depends on the researches now in progress by the Austrian palæontologists, that I forbear to enter upon a further consideration of the relations of that tract. Looking on the Alps as a whole, however, we have obtained sufficient proof that the lias and equivalents of the inferior oolite occupy a considerable thickness of the lower portion of what was called alpine limestone. This fact has been clearly established by the presence of fossils in the Venetian, Tyrolese and Milanese Alps. There are indeed tracts in which the *Gryphæa incurva* and several true liassic ammonites occur in rocks of this age, whilst in Switzerland and the Western Alps the zone has been traced and identified by Studer, Élie de Beaumont, Sismonda and many geologists.

In following these lower jurassic limestones from the Eastern into the Swiss and Western Alps, great changes in their mineral character are observable. In the first region they are very frequently light-coloured limestones often in the state of dolomite. In the west they are for the most part dark and even black. As, however, the chief strata in many parts of the Tyrol occur in the form of ordinary limestones, and as these can be followed in the strike until they are found to be transformed into dolomites, so this simple fact seems to me strikingly to corroborate the general view of M. von Buch, that the dolomites of the Alps have been produced by a modification or metamorphism of the original strata. Whatever may have been the proximate cause of this great metamorphosis—whether by certain hot vapours or gases, which rose from beneath during one of the revolutions which the chain has undergone, or by any other agent,—it is certain that this cause has acted not only vertically and obliquely, but in many instances horizontally over very large areas; thus transforming the superior strata and leaving comparatively unaltered those beneath. If the crystalline dolomites of the Eastern Alps were the result of original deposit like the magnesian limestone of England, as some geologists aver, then we should not see the irregular and, if I may so say, capricious diffusion of the dolomite, which far from affecting any one set of strata in their horizontal extension, is absent or present in rocks of various ages and at different horizons.

Whilst the great masses of dolomite are peculiar to the Eastern Alps, and are most striking in the tracts of the South Tyrol which have been penetrated by porphyries and other igneous rocks, their place is to a great extent taken in the Western Alps by copious masses of gypsum. In like manner, as the dolomite is the metamorphosed limestone, the lines of bedding and fossils of which are frequently left in the transformed masses, so are the great accumulations of gypsum apparently deposits of carbonate of lime which have been changed into the sulphate of lime. Whether on the highest part of the route over Mont Cenis, or in the deep gorges of the Tarentaise or the Maurienne, or in the Allée Blanche, the valley of Cormayeur and other tracts around Mont Blanc, the same lesson is invariably to be read off; viz. that great bands of limestone have been here and there, and often along zones of some length, converted into gypsum. For, the stratification and even the colours of the original mass so remain, the thick and thinly laminated beds of various tints of white and brown and grey limestone are so preserved, that I have frequently walked up to a rock under the persuasion it was a continuation of the limestone of an adjacent escarpment, until my hammer undeceived me. Whilst expressing my own belief, I must say that it is chiefly a development of the view entertained by the deceased Mr. Bakewell, who as early as 1820, when residing at the baths of Brida in the Tarentaise, came distinctly to the conclusion, that the great limestones of Savoy belonged to the upper secondary strata, and that the gypsum, whether anhydrous or granular, was subordinate to and interstratified with them. That author had also the merit of remarking how these subcrystalline rocks of secondary age were associated with talc schists and mica schists, and were all connected with the alterations due to the action of heat and the formation of granitic rocks*.

I will not here enter at greater length into the question of whether these masses of stratified gypsum, which sometimes occupy entire mountains, were contemporaneously deposited with carbonates of lime. If the presence of such thick accumulations of anhydrous gypsum be not due to metamorphism, I would ask those who entertain the opposite view to explain, how they can suppose that in the very same sea and at the very same period, carbonates of lime should have been deposited along many leagues of the sea-bottom, and that all at once the same laminæ of deposit should have been formed of sulphate of lime? How dovetail the one into the other? On the other hand, nothing is so natural as that the evolution of heat and gas in a region permeated by igneous rocks should have converted the original carbonates into sulphates in one tract, and have left the limestones unchanged in another.

Fortunately, indeed, nature still exhibits in the Alps of Savoy the process by which this conversion may have been effected. The well-known thermal waters of Aix, which rise from a great line of fissure, and contain a notable quantity of sulphur, do now actually change the ordinary jurassic limestone into the sulphate of lime, wherever their hot vapours charged with sulphuric acid have access

* See Bakewell's Travels in the Tarentaise, &c., vol. i. pp. 276 *et seq.* and pp. 289 to 311. Showing, according to M. Charpentier, that the granular gypsum of the Alps is simply the decomposed anhydrite, Mr. Bakewell points out the existence of a carbonaceous stratum in the heart of a thick band of gypsum.

to that rock*. We have only, therefore, to suppose, when some of those powerful changes occurred to which the Alps were subjected, that the more copious transmission of such hot springs and gases, operating on a grander scale and with much more intensity, produced commensurate changes of the carbonates into sulphates of lime, even throughout mountain masses, and also disseminated flakes of sulphur at intervals (as we find them) in the gypsum of the Alps. The conversion of ordinary limestone into the sulphate is usually accompanied in the Alps by other phænomena which forcibly bespeak metamorphism. The limestones situated near deep cracks and fissures (in some of which hot springs still exist, as at Moutiers) are frequently in the condition of a cellular hard tufa, sometimes siliceous, which is known in Savoy under the name of "Cargneule." No one can view this rock and not believe that it is the result of an action in which much heat and gas were evolved.

Although I have introduced the subject of these metamorphisms of limestone whilst speaking of the lias and lower oolite, I am by no means prepared to say, that the same transformation in the Western Alps has not also been applied to strata both of older and younger date ; just as limestones of different age in the Eastern Alps have been changed into dolomites. But however this may be, great masses of gypsum are certainly of about the age I speak of, inasmuch as belemnites and ammonites and other shells of the lias, including the *Gryphæa gigantea,* have been found in the associated strata at Bex by M. Lardi. Wherever certain ammonites, such as the *A. Walcotti,* and belemnites, or the *Gryphæa incurva, Plagiostoma gigantea,* &c., occur in the strata which occupy the lower zone of the alpine limestone, no one will dispute that such is about the horizon of the lias, whatever may be the mineralogical character or colour and structure of the rocks.

In some tracts saurians have recently been found in the limestones, as near Admont in the Austrian Alps, and at Perledo near Laico in the Milanese ; but the small and peculiar forms from the last-mentioned locality do not afford such sufficiently clear testimony concerning the age of the deposit†. I am not aware that any fishes of the liassic age have been detected in the Alps. M. Heckl of

* M. Joseph Bonjean, in an elaborate analysis of the mineral waters of Aix (Annales des Mines, vol. xvi. Third Series, p. 299), whilst treating of various effects produced by their acidulous vapours or gases, says—"Quelle que soit la nature des corps soumis à l'action de cette vapeur (gas acide sulfhydrique), ils sont tous rongés et détruits dans un espace de temps plus ou moins long. *C'est ainsi que les pierres calcaires, dont se composent les murs, se convertissent assez promptement en sulfate de chaux* à leur surface," p. 342. The intelligent physician of the baths, M. Despine fils, showed me the effects of this process, and gave me specimens in which limestone several inches thick had been so metamorphosed. This phænomenon of conversion of carbonate of lime into sulphate is also clearly described by Professor Mousson of Zurich, in a very able memoir on the district around Aix and Chambery. The subject of the occurrence of thermal waters along lines of dislocation is still more fully developed by this author in his sketch of the geology around Baden in Switzerland.

† See Memoir by M. Giuliano Curioni, Giornale dell' Instituto Lombardo delle Scienze, tom. xvi. p. 170.

Vienna, an excellent ichthyologist, showed me indeed a small round fish with a heterocerque tail from Perledo near Laico, the same tract in which the small saurian occurs, and which might lead to the belief in the existence of a still older deposit in that region. He also pointed out to me specimens of ichthyolites from bituminous schists associated with the limestones between Adelsberg and Zircknitz in Illyria, which are chiefly of the genera Lepidotus and Palæoniscus, and much resemble the fishes I formerly collected at Seefeld* in the North Tyrol, the strata of which, according to the present view of their succession, must be the equivalents of the lias and lower oolite. In Illyria, however, the above fishes are associated with a species of the genus *Thryssops*, which has not yet been found at Seefeld, but occurs at Solenhofen. These fishes indicate then the existence of jurassic rocks in that region.

In tracing the lias and Jura limestone through Savoy, I shall say little more than that, in following out the researches of M. Élie de Beaumont, Professor Sismonda has detected the presence of a sufficient number of fossils to characterise the liasso-jurassic and overlying Oxfordian groups, fossils of both of which are to be seen in the museums of Turin and Chambery.

Rocks containing Belemnites and Coal Plants in the Savoy Alps.—I may now allude to the much-agitated question of the plants of carboniferous species† being associated with belemnites. M. Élie de Beaumont and M. Sismonda contend that these plants (which are doubtless true carboniferous species) are interstratified with belemnites, notably at Petit Cœur in the Tarentaise, and that in many other parts of Savoy zones of similar plants occur, which are, in truth, prolongations of liassic or jurassic deposits, in which the well-known animal remains of those periods prevail. This opinion has met with antagonists, and the greater number of geologists, being naturally averse to what they consider an anomalous collocation, the recognition of which is attended with great difficulties, are disposed to receive with favour every effort which has been made to explain the phænomenon by reversal or plication. Until I visited the Savoy Alps, I confess that I was of this number; for the theoretical sections of M. Favre of Geneva, showing the possible curvature of beds the ends of which have been truncated, and the opinions at which other geologists had arrived, that a true representation of the carboniferous system existed in the Savoy Alps, and that the plants of Petit Cœur formed a part of it, had strongly predisposed me to coincide with such views. After an examination, however, of the case of Petit Cœur, I know not how to arrive at any other conclusion than that adopted by M. de Beaumont and M. Sismonda, the grounds for which I now proceed to explain.

* See Phil. Mag. and Ann. of Philosophy, 1829, vol. vi. p. 36. At that time, long before the days of Agassiz, I suggested that these fish might be of the age of the Thuringian schists. They are however clearly of liasso-jurassic age.

† Mr. Bunbury has recently shown that all the species of plants from these localities, examined by him at Turin, are true carboniferous forms—thus confirming the dictum of M. Adolphe Brongniart (Journ. Geol. Soc. vol. v. p. 130).

The chief natural feature which influenced me in relation to this question was, that wherever I turned my steps in the Savoy Alps, whether into the Maurienne or the Tarentaise, or towards the environs of Mont Blanc, I uniformly observed that beneath a zone distinguishable as lias by its fossils as well as by its intimate relations with overlying Jura deposits, there was (with the exception of a certain conglomerate and sandstone, often associated closely with such lias) no sufficient development to represent any of the inferior formations *from the trias inclusive downwards*; all the lower strata being in a metamorphosed and crystalline condition. In other words, it seemed to me that between the inferior crystalline rocks and the stratum with belemnites (often itself much altered), there was no adequate representative in space or time for the carboniferous rocks; which if ever they existed, must, therefore, I inferred, have been merged in the great metamorphosis which all the central portion had undergone. Nor could I avoid the query, if the schistose deposits in which true and beautifully preserved coal species occurred, were of the old carboniferous date, why no vestige of any palæozoic *animal* had been ever detected in the Western Alps, whilst in the Eastern Alps there are, as has been stated, animal remains of the triassic, carboniferous, Devonian and Silurian ages?

Let us now appeal to the facts of the section of Petit Cœur. On inspecting the map of Savoy, it will be seen that at Conflans or Albertville, the river Isère having hitherto flowed tranversely by Moutiers in a deep valley across the ridges of the Tarentaise, or from S.W. to N.E., makes a sudden bend and thence trends south-westward to Montmelian and Grenoble. This latter part of its course is, in fact, determined by mountains of jurassic limestone, which have the same general direction, and which constituting the outer zone of the Alps, are composed of the equivalents of the Oxfordian and upper oolites, together with the overlying cretaceous deposits hereafter to be alluded to. In ascending the transverse gorge of the Isère from Albertville to Moutiers, the geologist has no sooner quitted that outer calcareous zone and passed to the opposite side of the valley, than he finds himself immersed in talcose crystalline schists, in parts highly quartzose, and in parts having somewhat of a gneissose aspect. I do not pretend to describe every variety of these rocks, but it is worthy of remark, that in this narrow transverse gorge, whether they be talcose, micaceous, felspathic or quartzose, the strata in their central part appear to wrap over an ellipsoid of granite and granitic gneiss, which is in parts porphyritic*. As far as dip can be marked, these crystalline rocks between Albertville and the zone of granite are vertical; whilst to the S.E. or higher up they incline away from the granite. At all events, when the gorge widens and you approach to Petit Cœur, a village on the right bank of that river, the dip of the crystalline mass is decisively to the S.E. A coarse quartzose

* See the clear and copious account of all these phænomena in the narrative of the excursion of the geologists assembled at the meeting of Chambery in 1841, by the Abbé (now Canon) Chamousset, as well as M. Virlet's note on the porphyritic granite of La Batie (Bull. Soc. Géol. Fr. vol. i. New Series, pp. 166 *et seq.*).

conglomerate is here intercalated in the talc schists and dips with them. A little above or N. of Petit Cœur a mountain rill descending from N.E. to S.W., waters the highly inclined faces of these talc schists, or rather occupies a deep gorge in them, to the east of which these same beds are seen to graduate up into and to be surmounted by others which contain belemnites and plants, the whole in perfectly conformable apposition and at angles from 70° to 75°, as expressed

Fig. 2.

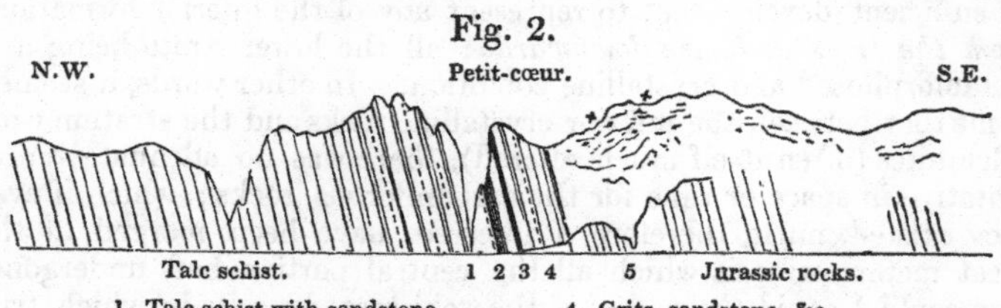

1. Talc schist with sandstone.
2. Black slates (Belemnites).
3. Talc schists with anthracite.
4. Grits, sandstone, &c.
5. Jurassic schists, &c.
†† Coarse detritus and blocks.

in the woodcut (fig. 2). The upper portion of the inferior or crystalline strata is a light-coloured talc schist used in the mountains for the covering of cottages. It is called "ardoise blanche" by the quarrymen, and forms the floor of the quarry (1). Immediately upon this, and perfectly parallel to it, lies another stratum (2) of fissile calcareous flagstone of dark indigo colour, called "ardoise noire," in which the belemnites occur. I could detect no slaty cleavage in these beds, whilst both the shining or talcose, and the dull black or calcareous flagstones strike N.N.E. and dip together 70° to 75° E.S.E. They are further connected by certain bands of hard sandstone occasionally coarse and gritty, with which the dark flags alternate towards their base. In the lower quarry (there being several openings in the line of strike in order to procure the black calcareous flagstone), these belemnitic beds are seen to be surmounted by another band of talc schist (3), as conformable to the belemnite flags or slates (2) as the latter are to the inferior talc schist (1). The uppermost of the three beds is that in which a certain portion of anthracite has been detected, the exploration of which has led to the discovery of numerous plants. These plants chiefly lie in the floor of the anthracitic schist, and therefore within a few feet of the belemnite flags (2), and they are strikingly distinguished by the brilliant white relief of the framework of the vegetables due to the dissemination of the talc upon the dark ground of the schist or flagstone. Much pyrites in single crystals and bunches occurs throughout the rock.

This carbonaceous schist with plants, several yards thick, is surmounted by strong beds of dark grey hard grit that weathers to a rusty colour, and which alternates several times with dark-coloured schists. And here it is to be observed, that the sandstone (4) above the coal plants is not to be distinguished from the band beneath the belemnite flagstone; or in other words, both the belemnite and the plant beds form parts of the same geological mass, the upper and lower parts of which are of similar composition, the talc schist and the sandstone being repeated. In fact, I cannot imagine how any geologist can look at this section and not declare that the whole of these strata form one natural group of very small dimensions. In tracing these beds up the hill-side I further convinced myself, that the belemnite flags (the belemnites most abound in the upper quarries) have there exactly the same relations to overlying and underlying grits, the whole resting, as in the lower quarries, on the white talc slates and also in perfect conformity. In these sections there can be no ambiguity, for you can absolutely follow the line of strike of each bed nearly a mile up the mountain-side. No traces of folding or contortion are observable ; and as belemnites have been found within a foot of the coal-plant bed, it appears, that however we may endeavour to explain them, the physical facts are clear and decisive.

It is true, that in ascending the hill the anthracite thins out from four feet to a few inches, whilst the black belemnite flags are more expanded in the higher than in the lower quarries ; but this phænomenon, so well known to every working geologist, is not worth mentioning, were it not necessary to allude to the minutest circumstance in this singular collocation. The whole group is affected by the same lines of joint or division, and all the beds exfoliate parallel to the laminæ of deposit, the more calcareous portions showing a tendency to assume flat concretionary forms, which produce small undulations in the upper quarries.

Above Petit Cœur the dip to the S.S.E. is continuous for a short distance, and there is therefore a certain amount of ascending order ; but in the parallel of Moutiers the succession is checked by one of those grand transverse dislocations so frequent in the Alps, accompanied by the evolution of hot springs.

According to M. de Beaumont and M. Sismonda, the belemnites, shales and flagstones of Petit Cœur are a portion of certain adjacent jurassic groups in which many ammonites and other fossils occur. These fossils are found in strata of dark shale and schist, which appear on the strike of these beds as they range across the valley of the Isère and appear in the passes which traverse the mountains that separate the Tarentaise from the Maurienne. It is probable that the same strata are several times repeated by fractures if not by undulations ; for it seemed clear to me that the whole of the series exposed between St. Michel and St. Jean de Maurienne, *i. e.* the dark shale and patches of coaly matter underlying or associated with jurassic rocks, represents the succession seen between Petit Cœur and Moutiers in the Tarentaise. The ammonites and other fossils I now present were taken from the broken slopes of the Col de la Madeleine, or rather from the elevated depression immediately to the S.S.W. of Aigues blanches, where they were collected by M. Ansenet. They appear to be in great part the same as those already collected in the tract at the Encombres by M. A. Sismonda, among which that author enumerates the *Ammonites fimbriatus*, Sow., *A. planicostatus*, Sow., *Avicula inæquivalvis*, Sow., *A. costata*, Sow., *Terebratula inæquivalvis*, Sow., *T. variabilis*, with a multitude of belemnites*.

* See a more complete list of these fossils, with specimens from other places in

If the previous statement of the geological relations at Petit Cœur be, as I think, accurate, *i. e.* if the plants and belemnites really lie in the same deposit, as was also concluded by the geologists of the French Society, who met at Chambery, the anomaly is great, and involves us in considerable natural-history difficulties. But are these difficulties insurmountable, and ought geologists to shrink from endeavouring to reconcile them because they interfere with the general distribution of fossil plants? Excluding for the present all theory, let me say that I cannot admit the presence of certain species of fossil plants to be as decisive of the age of a deposit as that of the remains of any well-known animal. Thus, the *Calamites arenarius* cited by Brongniart as pertaining to the old coal, is found both in the Permian system and in the Bunter sandstein and Keuper, or throughout the Trias, a system in which no one palæozoic animal has been detected. Again, the *Equisetum columnare,* which so abounds in the Brora (oolite) coal, and is most abundant on the Yorkshire coast beneath the Kelloway rock, is one of the most common of the trias plants of Germany. And yet as a whole, both the fauna and flora of the middle oolite and trias are utterly dissimilar. Look, on the other hand, to the weight attached to the presence of belemnites. In no instance has a belemnite been detected in any part of the world below the lias. The trias, of which there is not a trace in Savoy, but now so well known in the Eastern and Tyrolese Alps, affords no sign of a belemnite any more than the same group in other regions; still less has any one ever heard of a belemnite in an old carboniferous deposit in any part of the world.

In giving the previous description of the section at Petit Cœur, I have done so in opposition, I repeat, to the strong wish I entertained to be able to offer any explanation which might obviate the dilemma in which such a recognition places us. I tried, for example, to account for the phænomenon by an inverted dip, and endeavoured to reconcile the overlying position of the coal plants by a reversal, similar to that which clearly operated on the north face of Mont Blanc: but there the belemnitic strata plunge under crystalline rocks, whilst at Petit Cœur they overlie and are intercalated in them. I could not speculate on the crystalline rocks of the Isère being originally of younger age than the belemnite beds, like the well-known examples on the northern face of Mont Blanc in the vale of Chamonix, and having been metamorphosed by the influence of the ellipsoid of granite before adverted to; because if so, and that these fossiliferous beds were inverted, other older strata besides the mere bed with coal plants would be found above them, which is not the case; the same liassic or jurassic group being manifestly developed in considerable force on the line of dip.

Those geologists who have explored the environs of Mont Blanc have long been acquainted with the fact first indicated by Sir Henry de la Beche, that coal plants also appear in the well-known con-

Piedmont, which include other known species of Sowerby, V. Buch, Schlotheim, Agassiz, &c., as given by M. Sismonda, with a descriptive plate of the species from the Encombres (Bull. de la Soc. Géol. Fr. vol. v. New Series, p. 410).

glomerate of Valorsine, and it is fair to state that a very clear and instructive recent section of M. Favre indicates, that this band there forms (Col de Balme) the conformable base of all the liassic and jurassic deposits, whether altered or unaltered, of that highly disturbed tract*.

I do not wish, with my present knowledge, to press the question more closely. Those who have not examined the sections might theorize that the thin anthracitic zone of Petit Cœur was a mere shred, which had been left among the gorges of the pre-existing and crystalline rocks, but it is impossible to apply such a hypothesis to the case; for if the crystalline rocks on the Isère be of anterior date, then we see that the belemnites lie between them and the coal plants; and if they be altered lias and Jura, then it is almost incredible that a few feet of old carboniferous rocks should be so conformably interlaced with these younger deposits. It is just barely possible, that instead of the *vertical* truncated cone theoretically suggested by M. Favre to explain the anomaly†, the older carboniferous rocks may have here been thrown into a very rapid *inverted* anticlinal flexure, leaving a few feet only at their apex, and that jurassic or liassic strata have been conformably folded around this point, the whole having been since altered and denuded. But if so, it is certainly a section more deceptive than any I ever examined; and until I meet with other sites affording a different explanation, I can only repeat my belief, that the relations of the strata sustain the conclusions of M. É. de Beaumont‡.

Upper Alpine Limestone (*Oxfordian, &c.*).—I may remind English geologists, that the parallelism with their oolitic deposits which has been so elaborately worked out by M. P. Merian and other Swiss authorities in the Jura mountains, has, despite of change of mineral character and the much rarer occurrence of fossils, been successfully applied to the French and Savoy Alps by M. Sismonda, and to the Swiss Alps by Prof. Studer and M. Escher. But notwithstanding the former publications of Pasini and Catullo, the clear definition of an equivalent of the Oxfordian group, as established in the Savoy and Swiss Alps, had not been defined in the Southern Alps until M. von Buch demonstrated to the Italian geologists at the Milan meeting that their "Ammonitico rosso" was of Oxfordian age§. This view has since been much extended in respect to the Venetian Alps by M. de Zigno of Padua. In the excursion of the geologists of the Venetian meeting before alluded to, in the mountains of the Setti Communi, my friends and myself were convinced of the accuracy of the fossiliferous distinctions by which that geologist had separated the red ammonite limestone from the lower jurassic rocks on the one hand, and

* Bull. Soc. Géol. Fr. vol. v. p. 263.

† Remarques sur les Anthracites des Alpes, par Alphonse Favre. Tom. ix. Mém. Soc. Phys. et Hist. Nat. de Genève.

‡ The able Memoir of M. Scipion Gras, on the association of the carbonaceous deposits of the Isère, and on their passage into crystalline rocks beneath, and their being clearly separated from all liassic strata above them, is to be taken into consideration in settling this question. See Annales des Mines, vol. xvi. p. 361.

§ See Bull. Soc. Géol. Fr. vol. i. pp. 132 *et passim*.

from the overlying neocomian and cretaceous rocks on the other. In ascending from Pedescala, in the valley of the Astico, to the plateau of the Setti Communi, we passed first over the edges of a great mass chiefly in the state of dolomite, probably representing the lias and inferior oolite, and then over rubbly yellowish earthy limestone; next over other courses of greyish limestone containing Turritellæ, followed by limestone of deep red colour, which from the quantity of ammonites found in it has obtained the name of "Ammonitico rosso." This last is surmounted by the white neocomian limestone. In traversing the lofty plateau of the Setti Communi from Castel Bello by Rotzo and Roana*, we had numerous exhibitions of an ascending order, from grey earthy limestone, with some sandy beds, up into the same red ammonite limestones, and thence into the white neocomian, which is here a purely white indurated rock, and as much loaded with flints as our chalk of Antrim, which it much resembles. Without dwelling on any details, I will simply enumerate the succession of the strata of this tract, as proved from different sections which exhibit the strata more or less inclined, but all conformable, and in this ascending order :—

1. Dolomite, &c., of great thickness, probably representing lias. 2. Compact brecciated marls. 3. Beds of fine oolite, alternating with yellowish marly limestone, containing shells, including Diceras, Gryphæa, &c. 4. Thin-bedded limestone with Nerinæa and univalves. 5. Thin-bedded, dark grey, sandy limestone, with Neuropteris and other plants. 6. "Lumachello grigio," grey mottled limestone with marl, yellow marly beds and grey lumachello with sections of large bivalves. 7. "Ammonitico rosso." This rock is invariably the summit of all the jurassic strata in this region, and clearly represents the Oxford formation. Although seldom more than fifty or sixty feet thick, it is an excellent horizon, since it contains *Ammonites athleta* (Phill.), *A. anceps* (Reinecke), *A. Horneri* (D'Orb.), *A. Tatricus* (Pusch), *A. viator* (D'Orb.), with *Terebratula diphya* and *T. triangulata*. 8. "Biancone," or neocomian with *Crioceras Duvalli* (Leym.), *Belemnites latus* (Blainv.), *Ammonites asterianus*, *A. incertus* and *A. semistriatus* (D'Orb.), with some forms in the uppermost beds resembling those of gault, such as *A. Royerianus* (D'Orb.). In all, this group contains fifteen species of Ammonites, five or six species of Crioceras with Ancyloceras, Aptychus of two species, &c. 9. "Lower Scaglia," of grey colours with fucoids,—lower chalk. 10. "Upper Scaglia," of red and white colours,—upper chalk, &c. 11. Nummulite limestone and grit, with *Cerithium giganteum* and Nummulites of the same lower tertiary age as in the same position at Bassano; of which and of all the series from the neocomian upwards I shall treat hereafter.

In other parts of the South Tyrol, and notably at Trent, M. von

* This is the wild country of the ancient Cimri, and the people still talk a language unknown in any other part of the Alps.

Buch pointed out to me the clear determination of the Oxfordian formation by the presence of several of the fossils above enumerated; and in the grand natural sections between Rovereto and the Corneto pass, I also perceived a fine succession under the Ammonitico rosso.

The same Oxfordian zone has been delineated in Piedmont and Savoy by M. Sismonda. In the environs of Chambery and Aix les Bains, I had the advantage of studying its relation to the superior or cretaceous strata with the Canon Chamousset. That geologist has there divided the great Oxfordian masses into four parts, the lowest of which are marly limestones in beds of about a foot thick. 2ndly, Limestones, deep grey or bluish, with white veins and some ferruginous oolite. It is this band which contains most of the fossils, including the *Ammonites biplex*, &c., with Aptychi, &c. 3rdly, Foliated marls or calcareous flags; and, 4thly, Marly limestone, &c. This zone is overlaid by limestone with many corals, which is paralleled with the coral rag of English geologists, and that again is conformably overlaid by a full expanse of the neocomian formation, which in this region is divisible into three parts.

Not describing all these strata, I here merely annex a general section from the base of the Oxfordian to the summit of the Neocomian, which I made on the western shore of the Lac de Bourget on the sides of the zigzag road from Chambery to Lyons, which traverses the Montagne du Chat†. This Oxfordian group forms the

Fig. 3.

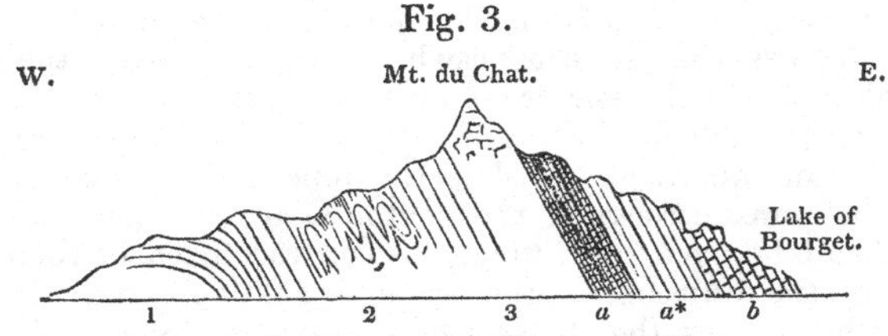

Cretaceous. { b. Upper Neocomian limestone with *Chama ammonia*.
a*. Middle Neocomian with *Spatangus retusus*.
a. Lower Neocomian, greenish ferruginous calc grit with *Ostreæ*, *Pectens*, &c. = base of English Lower greensand.

Jurassic. { 3. Coralline limestone = Coral rag. Summits dolomitic.
1. Oxfordian Jura with Kelloway rock fossils. 2. Schists and limestones.

base of all the outer edges of the Savoy Alps; their summits usually consisting of neocomian limestone, and often covered by still younger rocks. This order is seen around Chambery and the lake of Bourget at Annecy, in the valley of the Arve, both above and below Sallenches, and near Geneva.

The Oxfordian limestones, but without such a capping, are copiously exhibited in the range of mountains east of Vevey. At Chatel St. Denis, where the limestone is very mottled and concretionary, it is loaded with ammonites and Aptychi‡; and the same rock, subjected to great flexures, rises into the high peak, the Dent

† See Bull. Geol. Soc. Fr. new ser. t. i. pl. 9. fig. 3.

‡ I was directed to these limestones by M. Studer, and M. Collon of Vevey accompanied me thither.

de Jaman*, so well known to tourists, where masses of the mottled limestone with flint nodules repose on dark schists and impure limestones as thus represented. From that peak there is apparently a

Fig. 4.

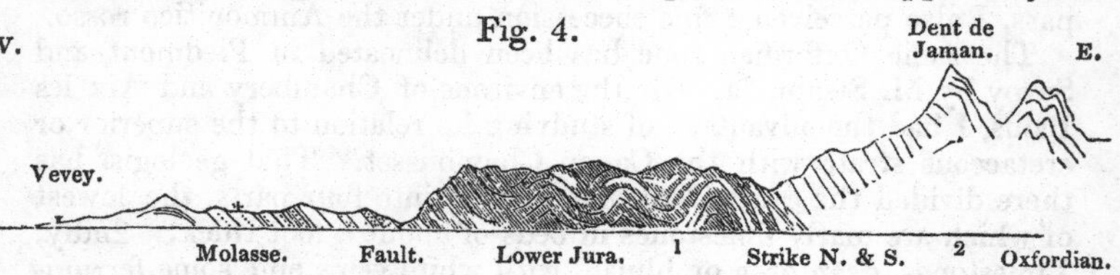

descending succession through other limestones, as exposed in the gorge of the Baye de Montreux near Glion, down to the black fetid limestones and shale at the bridge of Montreux, which may represent the lias†. Again, when wandering through the Swiss Alps, whether on the lakes of Thun and Brientz, on the upper portion of the Lake of the Four Cantons (fig. 12), or on the northern shore of the lake of Wallenstadt (fig. 14), I came upon calcareous bands, which Studer, Escher, and the best Swiss geologists consider to be of Oxfordian age, sometimes surmounted by limestones the equivalent of the coral rag, and sometimes without them, but invariably covered by the neocomian limestone, as in the Savoy section (fig. 3). The Portland limestone, so copiously developed in the Jura, and so rich in fossils at Soleure, has not as yet, to my knowledge, been found in the Alps‡. It is only indeed by fossils (and unfortunately they are of rare occurrence in the alpine limestones) that the strata can be actually referred to the respective members of the jurassic or oolitic series. Resting upon carbonaceous schists, which in their turn overlie the so-called "Sernft conglomerate" and quartzose talc slate, the upper part of the lower division of the jurassic rocks of the canton Glarus, in parts dolomitic ("Zwischen-bildungen" of Studer), are characterized in their upper member by the diffusion of iron ore in an oolitic matrix. It is this rock which contains the *Ammonites Gowerianus* (Sow.), *A. macrocephalus* (Schloth.), *A. Parkinsonii* (??), *Ostrea pectiniformis* (Schloth.), *O. calceola* (Goldf.), and *Terebratula digona* (Sow.), &c.

The overlying stage, or that which immediately succeeds to the ferruginous oolite (the "Hoch-Gebirg's Kalk" of the late M. Escher), as seen in the cantons of Glarus and Appenzell, contains the characteristic forms *Ammonites biplex* and *A. polyplocus*, with belemnites, and is therefore a good representative of the Oxfordian of the Alps§.

* 1872 metres above the sea.

† M. Collon informed me that the *Ammonites Petit Thouars* (d'Orb.), a lias fossil, had been found in the black limestones and schists which are exposed in the gorge above the little bridge, dipping under the mass of the adjacent mountains.

‡ The abundance of tortoises and other peculiar fossils which characterize the "Portlandian" of Soleure, indicate the *local character* of the formation, and the same may, indeed, even be said of the Portland rock of England.

§ For the description of the lithological varieties of these Alpine jurassic rocks, see the Gebirgskunde of M. Arnold v. Escher, included in the general natural-history account of the Canton Glarus by Professor Heer of Zurich.

CRETACEOUS SYSTEM, *composed of Neocomian Limestones = Lower Greensand; Gault; Upper Greensand, and Inoceramus Limestone or Chalk.*—In noting some features of the jurassic or oolitic rocks, as traceable through the Alps, I have already pointed out several natural sections which show, that the rocks which in the ascending order are the equivalents of our oolitic series are conformably surmounted by other limestones, the "Neocomian" of foreign geologists*. In England, as I anticipated it would prove to be, and as we now know through the labours of Dr. Fitton and others, our lower greensand, if not the exact equivalent, represents a large portion, at least, of the neocomian. In the Alps this formation is so linked on to the alpine limestone, that before it was distinguished by fossils, Professor Sedgwick and myself, considering it simply the uppermost member of the great calcareous mass of the Alps, referred it with the geologists of that day to the upper oolite. Our stratigraphical view is, indeed, even now quite correct; for, with a few local exceptions cited by other authors, it seems that in the Alps, as in the Jura, there has been a continuous series of marine deposits in which no general disseverment took place, until after the completion of the supracretaceous nummulitic group (see figs. 12 and 14 in subsequent pages). M. Favre has, it is true, endeavoured recently to show, that in the Alpine tracts around Mont Blanc, the cretaceous system (*i.e.* from the neocomian up to the nummulitic zone inclusive) occurs in more or less horizontal bands, which rest on the convoluted strata of the jurassic age†. It is not in my power to controvert the specific cases which that geologist has cited; but other evidences will presently demonstrate, that even in the same region there are many proofs of the uninterrupted and conformable succession I have spoken of, and which is so clearly seen in the Venetian Alps. No one who has examined the mountains near Chambery in Savoy, or the flexures and contortions to which the whole of the secondary series has been subjected in the little Cantons of Switzerland, and who has seen the manner in which even the supracretaceous as well as the cretaceous beds fold over and conform to the convolutions of the jurassic rocks beneath them, could, I think, hesitate in adopting the conclusion at which I have arrived.

Not, however, to anticipate what I wish to demonstrate by evidence, I may in the mean time say a few words on the general structure and characteristic features of the Alpine cretaceous system properly so called. Its lower member, the neocomian limestone, is by far the thickest and most important cretaceous formation. This deposit has already been adverted to in the Venetian Alps as a hard white limestone with many bands and geodes of flint, and numerous characteristic fossils; and it there dips under the grey, red and white scaglia or chalk. In the Austrian Alps it is the hippuritic lime-

* See my observations on the equivalents of the neocomian at the Meeting of the French Geological Society at Boulogne, anno 1839 (Bull. Soc. Géol. Fr. vol. x. p. 392), and my Address to the Geological Society of London, anno 1843 (Proceedings Geol. Soc. Lond. vol. iv. p. 112). I was not aware, at that time, that Captain Ibbetson had expressed the same opinion at Neufchatel.

† See Bull. Géol. Soc. Fr. vol. iv. p. 996.

stone and marble, of grey, yellowish, and occasionally pink colours, which at Untersberg near Salzburg, around the valley of Gosau, and at numerous other places, plunges under strata of impure limestone, marl and sandstone, charged with fossils of the gault and chalk. In the western parts of Savoy, however, it is more clearly divisible into three zones than at the city of Neufchatel itself, and the diagram of the Montagne du Chat already given, explains the fact (see fig. 3).

In another section west of Chambery, which I made in company with the Canon Chamousset, the order of the strata is exhibited in this diagram (fig. 5). The lowest rocks visible are the Oxfordian

E. by S. Fig. 5. W. by N.

b a* a 1 & 2 a a* b m m* m b a* a 1 & 2

Tertiary. { m*. Marine molasse.
m. Freshwater limestone and conglomerate.

Cretaceous. { b. Upper Neocomian (*Chama ammonia*).
a*. Middle Neocomian (*Spatangus retusus*).
a. Lower Neocomian (*Oyster bands*).

1 & 2. Oxfordian Jura and coral rag limestone.

limestones and shale, and the coralline limestone before adverted to (Nos. 1 and 2) which constitute the uppermost jurassic band of this region. On this reposes the lowest neocomian (*a*), which is a hard siliceous limestone with small, sharp-plaited Ostreæ, a small Terebratula, Nerinæa, &c. The middle neocomian (*a**) consists of alternations of bluish grey marly limestone and bands of green-grained calc grit and beds of chert, and in this band most of the fossils occur, including the very characteristic form *Spatangus retusus*. The upper division (*b*) is a whitish limestone, often in a state of marble, which in Savoy contains both Hippurites and the *Chama* (*Caprotina*) *ammonia* (D'Orb.).

For the most part, however, in its prolongation along the flanks of the Savoy, and particularly in the Swiss Alps, the neocomian is divisible into two great subformations only; the lower being dark-coloured and marked by the *Gryphæa Couloni* (Leym.), ***Rhynconella*** (*Terebratula*) *depressa* (D'Orb.), and *Spatangus retusus* (Lamk.) (Spatangus-kalk, Studer), and the upper being a light-coloured limestone containing the *Caprotina ammonia* (the Schratten-kalk of Escher), is a sure and excellent horizon throughout the greater part of the Alps†.

Cretaceous Greensand or Gault of the Alps. (*Turriliten-Etage*, Escher.)—The largely exposed neocomian limestone of the Savoy Alps supports, as above stated, in various escarpments, a thin zone of dark-coloured marly limestone, occasionally freckled with grains of chlorite, and abounding in fossils. In a collection made by my guide, Auguste Balmat of Chamonix, at the Montagne des Fis, Professor Pictet of Geneva recognized *Ammonites cristatus* (De Luc); *A. Hugardianus* (D'Orb.); *A. Mayorianus* (D'Orb.); *A. inflatus* (Sow.);

† It is worthy of note, that this upper band, as distinguished by the *Caprotina ammonia*, is absent at Neufchatel, as well as the lower part of the formation.

A. splendens? ; *Hamites alternatus* (Sow.) ; *Nautilus,* small species ; *Avellana incrassata* (D'Orb.) ; *Inoceramus sulcatus* (Sow.) ; *Solarium ornatum*? (Sow.) ; and a new species, together with various *Echinidæ* (*Discoidea, Galerites* and *Micraster,* Ag.).

In laying these fossils before the Society I also present a certain number from Sassonet, near Bonneville, and the Reposoir, which Professor Pictet kindly gave to me. The mere view of these fossils will convince English geologists that the rock of which I am now speaking fairly represents their gault and upper greensand*. A band of this age which I shall indicate in other natural sections in the Swiss and Bavarian Alps containing some of these characteristic fossils, is at intervals traceable far into the recesses of the higher mountains.

Inoceramus Limestone (*Sewer-kalk*), *equivalent of the chalk of Northern Europe.*—When I visited the Savoy Alps, it was still to be ascertained whether they contained any equivalent of the white chalk of Northern Europe, which surmounting the upper greensand was there fairly intercalated between that formation and the great "Terrain à Nummulites." In entering that region last summer I was indeed led to believe, from the first sections I observed around Chambery, that there was little chance of meeting with so full a succession of all the cretaceous strata as would exhibit any equivalent of the white chalk, for there the nummulitic rocks, as above stated, repose at once, as pointed out to me by the Canon Chamousset, on neocomian limestone. Moreover, in his very last memoir†, Prof. Favre had described the nummulitic zone in Savoy as independent of the cretaceous system on the one hand, and of the overlying macigno or flysch on the other. That geologist had doubtless reasons for such an inference, in seeing that the nummulitic rocks, where he examined them, reposed in one place on Jura limestone and at another on neocomian ; but such reasoning in a region which has been subjected to many dislocations, is liable to be overturned by the discovery in an unbroken tract of the beds supposed to be wanting. Such, in truth, I found to be the case in a very clear natural section exposed at Thones in Savoy, which I examined in company with M. Pillet of Chambery in a traverse from Annecy by the valley of Thones and the Grand Bornand to the Col du Reposoir, and thence to the valley of the Arve.

In entering the valley of Thones from the west, I perceived, near

* The fossils given to me by Prof. Pictet from the above localities and from the Perte du Rhone, where the same species occur,—in all three places usually in a bed of a few feet thick only,—are *A. inflatus* (Sow.), *A. Candollianus* (Pict.), *A. varicosus* (Sow.), *A. Mayorianus* (D'Orb.), *A. Lyellii* (D'Orb.), *A. monile* (Sow.), *A. milletianus* (D'Orb.), *A. regularis* (Leym.), *A. latidorsatus* (Michelin), *A. Hugardianus* (D'Orb.), *Hamites rotundus* (Sow.), *H. virgulatus* (Brongn.), *Turrilites Bergeri* (Brongn.), *Avellana incrassata* (D'Orb.), *Inoceramus concentricus* (Sow.), *I. sulcatus* (Sow.), *Cucullæa fibrosa* (D'Orb.), *Arca,* three species, *Terebratula ornithocephala* (Sow.), *T. plicatilis* (Sow.), *Ceromya inflata* (Ag.), with *Micraster* and other Echinoderms. See M. Pictet's excellent work, "Description des Mollusques Fossiles des grès verts des environs de Genève," 1re livr., 1847.

† "Sur la position relative des Alpes Suisses occidentales et des Alpes de la Savoie."—Bull. de la Soc. Géol. Fr. vol. iv. p. 996.

Annecy le Vieux, an ascending succession from neocomian to overlying nummulitic rocks with Pectens; but the broken nature of the banks at this locality, and the little time at my disposal, prevented the tracing of the intermediate strata. On approaching the village of Thones, it was however seen, that after several flexures the upper strata of neocomian limestone with *Caprotina ammonia,* forming a striking ridge on the north side of the valley, and having here a north and south strike, plunged south-east at an angle of 55° to 60°, throwing off on its surface the other strata exhibited in this diagram.

Fig. 6.

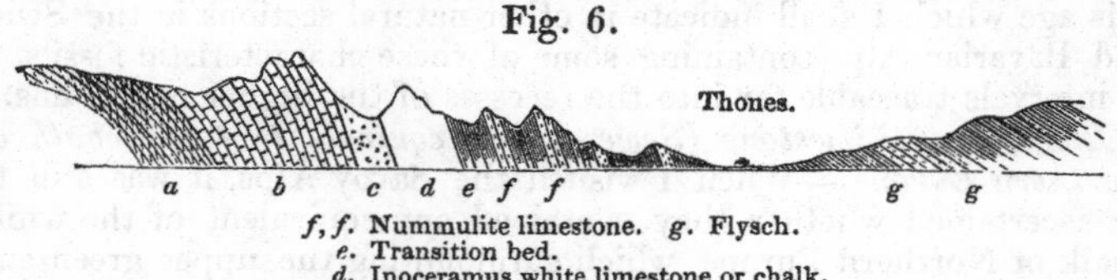

f, f. Nummulite limestone. *g.* Flysch.
e. Transition bed.
d. Inoceramus white limestone or chalk.
c. Gault and greensand.
a & b. Neocomian limestones.

On ascending to the little depression above the surface of the upper neocomian (*b*), it was evident that the excavation was due to the beds being less coherent than the hard limestone (*b & d*) above and below them. M. Pillet and myself then detected greensand terebratulæ, which, when combined with the position and character of the strata, led me to believe that this band of dark shale, impure limestone, and sandy green marlstone (*c*) was the representative of the gault and upper greensand usually exhibited in these Alps. These beds graduate upwards into yellowish limestone, which is surmounted by a cream-coloured compact limestone with flints weathering white (*d*), in which we found several specimens of Inocerami, the best-preserved of which was the *I. Cuvieri.*

Here then we had under our feet a band (*d*) which by position and fossils must fairly stand for the chalk. In proceeding upon the dip this limestone is seen gradually to change its colour from white to brown (*e*), and in a short space, without the slightest break or unconformity of the strata, the overlying mass is charged with nummulites. The nummulitic strata (*f*) becoming sandier upwards, contain also certain Pectens, and these shelly beds are overlaid by a zone of coralline concretionary sandy limestone, and the latter by a strong-bedded, dark grey, white-veined limestone. This nummulitic group, so intimately linked on to the Inoceramus limestone (or chalk) in its lower beds, is quite as intimately connected with the strata by which it is overlaid, *i. e.* with micaceous sandstones, marls, impure limestones and conglomerates, which represent a portion of the "flysch of the Alps (*g*)." These flysch beds contain certain scales of fishes enumerated by Agassiz, and some casts of shells. They are, in short, identical with the strata which at the desert near Chambery had been pointed out to me as the recognized lower beds of the "flysch" of Savoy, where they equally overlie and are equally connected with bands of nummulite limestone.

In ascending the lateral longitudinal valley of the Borne, which extends eastwards to Grand Bornand, particularly between Thones and St. Jean de Sixt, you have on the left hand a ridge of neocomian and cretaceous limestones overlapped by nummulitic limestone with some of the schist of the flysch, and on the other side of the road superior strata of the same series with hard sandstones in which conglomerates appear. There is no interval except that which has been occasioned by denudation, and all the strata are conformable and highly inclined, dipping to the S.S.E., the angle of inclination decreasing with the distance from the secondary limestones. At La Sommerie, to the east of Grand Bornand and in a deep depression under the Montagne de Four, a lignite has been partially worked, of the same age as that of Entrevernes near Annecy, which is fairly intercalated in the nummulitic group*.

It is therefore evident, that even in the environs of Mont Blanc, there is a connected section, which not only exhibits the whole succession of the cretaceous rocks properly so called, but also their upward lithological transition into beds with nummulites; and further, that the latter are inseparable from the overlying flysch. The independence, therefore, suggested by M. Favre does not exist in this part of Savoy where the natural original relations have not been effaced by dislocations.

Now, these nummulitic and "flysch" strata, which by much more developed natural sections in Switzerland, as well as by a consideration of their fossils, will be proved to be a natural group, distinct from, yet intimately and conformably linked on to the cretaceous system, are copiously exhibited on the summits of some of the highest and least accessible of the calcareous mountains to the north-east, north, and west of Mont Blanc. Thus, rising to vast altitudes, they cap the Dent du Midi and Diableretz, the former 9849, the latter 10,050 French feet above the sea. The fossils of the summit of the latter have been long known to geologists, and besides *Nummulites globulus* (Leym.), the *N. Biaritzana* (D'Arch.) or *regularis* (Rüttimeyer), are the *Cerithium diaboli* (Brongn.), *C. elegans* (Desh.), *C. polymeres* (Leym.), together with *Ampullaria*, *Chemnitzia*, and the *Melania costellata* (Lamk.), three of which are undistinguishable from species of the Paris basin.

In his admirable description of the rocks composing the summit of the Diableretz, M. Brongniart not only enumerated nummulites and several other fossils, and also indicated the intercalation with them of a band of combustible in the condition of anthracite, but he further justly reasoned on the nature of the shells and on the

* This coal of Entrevernes is noticed by Bakewell, Travels in the Tarentaise, vol. iv. p. 186, with woodcut. This author mentions Cythereæ and Cerithia, but does not allude to Nummulites. It was also visited by the members of the meeting of the Geological Society of France which assembled at Chambery, when MM. Chamousset, De Verneuil, Sismonda and Viquenel are reported to have found tertiary shells associated with it. See Bull. Soc. Géol. Fr. 2nd series, vol. i. p. 214. Coal of this age also occurs in the summit of the Diableretz (see next page) and at Pernant on the Arve, where it was observed by Prof. Necker in both situations associated with nummulites. For the latter position see Bibl. Un. de Genève, tom. xxxiii. p. 90.

whole of the evidences as disposing him to view these rocks as being of about the same age as those of the lower strata of the Paris basin. He well distinguished the nummulitic and carbonaceous black limestones from those of the adjacent mountains in which Ammonites, Hamites and other secondary greensand fossils occurred, and was only disposed to doubt his conclusions by the very ancient lithological aspect of the overlying schists and limestones. Such was the influence of mineral character in those days *! Now, however, that representatives of every band of the upper secondary or cretaceous rocks are known to exist in these Alps of Savoy and the Vallais, including even the equivalent of the chalk, we see how sound were the first conclusions of M. Brongniart as to the true tertiary age of the black nummulitic limestones of the Diableretz.

Intending to explore the relations of these supracretaceous strata in the Swiss Alps, where the labours of the geologists Studer and Escher de Linth had already succeeded in developing to a great extent their order, I abandoned further researches on this point in Savoy and the Vallais, being satisfied with having there detected a key to the order of superposition which had escaped previous observers. I further presumed that the limestone with Inocerami, which I had there observed to be intercalated between the greensand and the nummulite rocks of Savoy, would prove to be the same as the Sewerkalk of the Swiss geologists, and future researches completely established this to be the fact.

Nummulitic Rocks and Flysch of Switzerland ("Macigno Alpin" of Studer), with their relations to the subjacent cretaceous rocks.

Having touched upon the cretaceous and overlying masses of the Savoy Alps, I now proceed to describe in greater detail a series of sections specially illustrative of the sections I made chiefly either in the company of Professor Brünner of Berne or in that of M. Escher de Linth of Zurich, in the cantons of Lucerne, Underwald, Schwyz, Glarus, Appenzell and St. Gallen. In so doing I shall necessarily often refer to the underlying cretaceous rocks. As the general view of succession has been already given, it is deemed more desirable, for the better understanding of the subject, that the whole series of strata in each tract which are physically connected with the nummulitic zone should be collectively described, rather than first enumerate all the cretaceous rocks in different districts and then revert many times to the same place to describe the supracretaceous deposits. This would entirely frustrate my object of showing in consecutive sections the inti-

* "J'hésiterais donc très peu (says M. Brongniart), malgré la position de la roche calcaire qui renferme ces fossiles, malgré sa compacité, sa couleur noire, sa stratification concordant avec le calcaire ancien qui est au dessous; j'hésiterais peu, dis-je, à la regarder comme de même formation que le calcaire grossier de sédiment supérieur, si elle n'était recouverte par des roches qui offrent de nouveau le caractère d'homogénéité et de compacité qu'on attribue au calcaire alpin," &c. (Mémoire sur les Terrains de Sédiment supérieur, p. 44.) These overlying blackish, siliceous and micaceous sublamellar impure limestones, and compact scaly limestones with white veins, are parts of the "flysch."

mate connexion of some rocks which must, I conceive, be considered tertiary, with others which are unquestionably of secondary age.

And here geologists will recollect, that when Professor Sedgwick and myself wrote upon the Austrian Alps, the structure of the interior and flanks of the Swiss Alps had not been illustrated by Studer, Escher, and others. M. Studer had then, it is true, published a portion of his excellent work on the Molasse, but his first attempt at a classification of the older formations* had not appeared. Good even as that effort then was, it now requires much revision to bring it up to the state of our present knowledge; and so must it ever be in so complicated and difficult a chain†.

On attending the meeting of the Swiss naturalists at Soleure (after I had passed through Savoy), I was fortunate enough to hear a memoir read on Nummulites and other Foraminifera by M. Rüttimeyer of Berne. On a previous occasion Professor Brünner had described some of these forms as well as their geological position‡; but desirous that the purely zoological portion of this labour should be undertaken by a professed naturalist, he engaged his friend M. Rüttimeyer to join him, and the first result was the memoir I have alluded to, which will be followed by the publication of a joint work. In pursuing my inquiries I induced Prof. Brünner to accompany me in excursions into parts of the little cantons which he had not explored.

In the Beattenberg near Thun a band of coal is associated with the nummulitic deposit, *i. e.* in the strata beneath the flysch. This coal, which is now extensively used in the manufacture of gas at Berne, is therefore precisely in the same geological horizon as the coal of Entrevernes near Annecy, of Grand Bornand in Savoy, and of the Diableretz. In this respect there is indeed a close analogy between the northern and southern flanks of the Alps; for, as will hereafter be shown, coal is pretty largely extracted from the lower strata of the nummulitic rocks of the Vicentine, between Vicenza and Recoaro, and at Monte Bolca, in a region where these deposits unquestionably overlie everything cretaceous.

The nummulitic rocks of the Beattenberg (*f*) at once repose on the neocomian limestones (*b*), and are surmounted by flysch (*g*),

* See Transactions of the Geological Society of France, 1834.

† No one is more aware than M. Studer of the necessity of frequent revisions and corrections of all the older sketches or attempts to map geologically any portion of the Alps before the organic remains were developed. In reference to his own small map of the region around Berne, he candidly explained to me, that the legend attached to it must now be much changed.—See Trans. Geol. Soc. Fr. vol. iii. p. 379.

‡ See Professor Brünner's memoir, "Beiträge zur Kenntniss der Flysch und Nummuliten Formation," Mittheilungen der Naturforschenden Gesellschaft zu Bern, 1847. In this memoir Professor Brünner compares the nummulitic strata to the north of the lake of Thun with those of the Diableretz, the *Nummulites globulus* (Leym.) being common to both. In both are species of Cerithia, Chemnitzia, &c., whilst the *Neritina Fischeri* (Brünner) of the Thun district is scarcely to be distinguished from the *N. lineolata* (Deshayes) of the Paris basin.

M. Rüttimeyer has since published an extract from his work in the Bibliothèque Universelle de Genève.

(fig. 7). In the countries, however, to which Professor Brünner and myself extended our researches, we perceived the relations of the de-

Fig. 7 (*by Professor Brünner*).

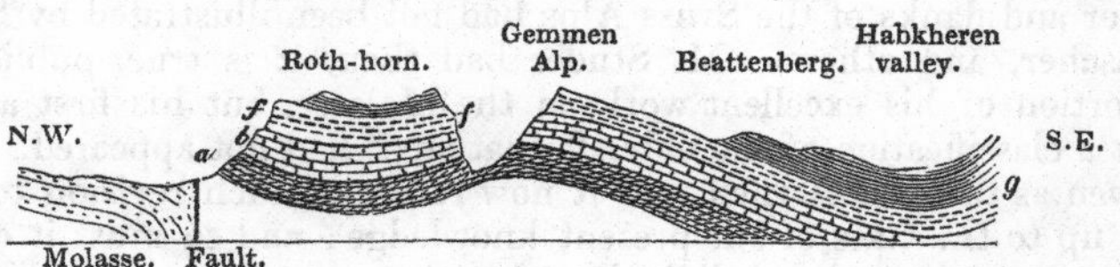

posit to the subjacent rocks to be much more copious and clear than what he had previously known, and more in harmony with my section at Thones in Savoy. Finally, meeting M. Escher de Linth, I found that this excellent field-geologist had (with the exception of a peculiar band of passage on which I lay great stress) come to the same conclusion as myself concerning the true position of the nummulitic zone, as being invariably above the inoceramus limestone or representative of the chalk.

The great zone of limestone, containing Nummulites, Orbitolites and Operculinæ, with certain shells, and surmounted by vast accumulations of "flysch," *i.e.* impure limestone, sandstone and schist, extends from the Beattenberg and Habkheren to Alpnach, trending parallel to the major axis of the Swiss Alps, viz. from W.S.W. to E.N.E. It is, in fact, an elevated trough between the great calcareous chains of Hofgant, Sernberg and Pilatus on the N.W., and the ridges which flank the lakes of Brientz and Sarnen, the Stanzhorn forming the south-eastern "pendant" to the Mount Pilatus. The depression occupied by the Alpnach branch of the Lake of the four cantons has been essentially formed in the softer schists or marly shale and sandstones of the "flysch" deposits, whilst the hard calcareous rocks on the flanks of the trough constitute the Pilatus on the one hand and the Stanzhorn on the other. I do not pretend to have so examined Mount Pilatus as to be able to give a detailed description of its structure and relations. I ascended it from Alpnach to the south flank of the Thumli-horn, and thence by a valley, leaving the Eck-horn on the right hand, to the summit, called the Esel, about 6000 feet above the sea, devoting the short time at my disposal to the examination of the nummulitic strata and the rocks on which they rest. The chief masses of the mountain are certainly composed of upper neocomian limestone (with *Caprotina ammonia*); and between the Thumli-horn on the one hand, and the Rustiger-wald on the other, I perceived a brownish sandy limestone which strongly contrasted with the white neocomian limestones of the flanking mountains. I saw no traces of gault, upper greensand, or inoceramus limestone, but judging from the analogies on the eastern shore of the lake of Lucerne, hereafter to be described, it is probable that such may be found on one of the unbroken shoulders (if such there be) of this remarkably bold and highly dislocated mountain. From a spot near the Esel summit, where I observed nummulites*, I perceived that there was an ascending section, with a rapid dip to the S.S.E., through beds of impure limestone into highly ferruginous strata, which in parts became a strikingly green calcareous grit (in parts small pisolitic), in which were casts of Pectens and other shells, similar to those associated with nummulites in many other parts of Switzerland. These green sandstones and calc grits there dipped rapidly under a vast thickness of schists, micaceous sandstones and bastard limestones; in short, under the "flysch." It was thus clear that the nummulitic and flysch rocks, though perfectly united and conformable within themselves, and clearly forming one natural division, were at this high gorge unconformably enclosed between two walls of the older neocomian limestone, as exhibited in this diagram. In my rapid survey

Fig. 8.

Gorge east of Mt. Pilatus.

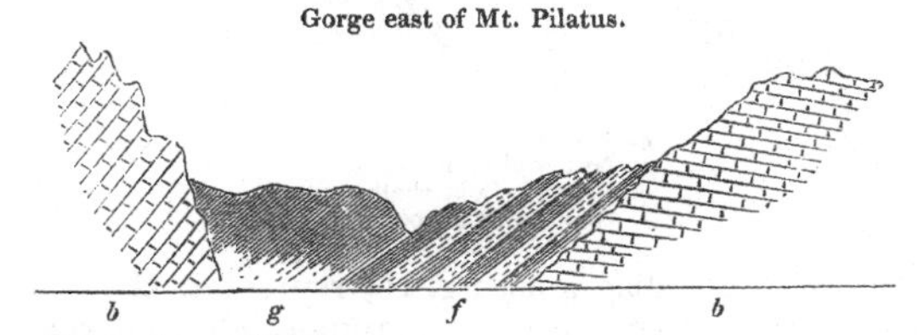

g. Flysch of great thickness.
f. Ferruginous greensand with Pectens (part of the nummulitic group).
b. Neocomian limestone (upper).

I did not visit the adjacent flanks of the mountain in which a sequence might be found; and I have only to observe, that in the great masses of finely laminated marly and sandy schists which descend rapidly on the face of the older limestones into the lake, I found some of the same small foraminifera which MM. Brünner and Rüttimeyer have recognized in the environs of Thun.

On the whole, however, the nummulitic and flysch rocks of the Pilatus have the appearance of having been upheaved in a highly broken and elevated trough, the sides of which rest on the edges of the neocomian limestone, which latter presents to the north one of the finest mural precipices along the whole outer edge of the Alps, to the lower and undulating country of molasse and nagelflue, which here range over the canton of Lucerne.

The eastern end of the lake of Alpnach is almost barred in by a tongue of land, composed of subconical and undulating hills, which

* The species of nummulite I found in the Pilatus was small, but it is well known that large forms of this genus are there also present. In reference to these organic remains, I ascertained, when in the company of Professor Brünner, how much the species of Nummulites and other Foraminifera differ in the same region at different localities, and yet, as will hereafter be seen, the very same characteristic species reappear at spots very widely distant from each other.

having been examined by M. Brünner, proved to form a very instructive trough, as expressed in figs. 9 and 10, the lowest rocks on either side being the upper neocomian (*b*), surmounted by the Sewer-kalk (*d*) or equivalent of the inoceramus limestone, and this by a basin of nummulitic limestone and flysch (*f*, *g*)*.

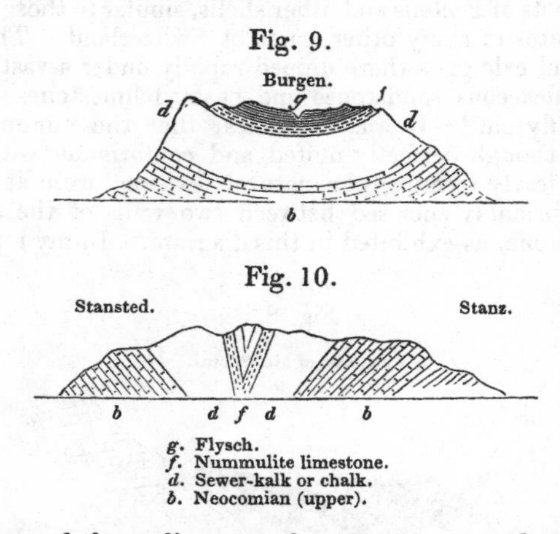

g. Flysch.
f. Nummulite limestone.
d. Sewer-kalk or chalk.
b. Neocomian (upper).

The lowest of these diagrams, fig. 10, represents the general relations at the south-western end of the promontory, between Stansted and Stanz, where the nummulitic rocks are squeezed up, whilst fig. 9, on the strike of the same strata to the W.N.W., shows how the basin of nummulitic and flysch rocks expands and becomes regular.

The Orbitolites, which occupy beds of considerable thickness in the mountains of Ralligstock and Beattenberg, near Thun, are here contained in a green calcareous sandstone of a few feet thickness only, whilst the *Nummulites millecaput* (Boubée), or *polygyrata* (Desh.), is much developed, and seeming, according to Brünner, to replace in this spot the small *N. rotularis* (Desh.), or *N. globulus* (Leym.). The first-mentioned large and striking species, which is so extensively distributed over the globe, reappears in many other tracts to the north-east, as will be hereafter detailed†.

* When M. Brünner examined this promontory he had not had his attention called to the thin band of secondary greensand or gault which we afterwards found so usually intercalated (as in Savoy) between the upper neocomian and the inoceramus limestones; and in a rapid examination, looking chiefly to the great relations and general symmetry of the trough, a few feet of greensand may he thinks have escaped him. (See fig. 10.)

† In my tour I necessarily used the specific names given to the Nummulites and other Foraminifera of the Swiss Alps by Rüttimeyer and Brünner; but on comparing the forms I collected, M. D'Archiac, to whom I referred them, identifies several of them with species previously named and described in France. Thus, whether the following names, as given in Italics, be finally adopted or not, their equivalents being here mentioned, no misunderstanding can arise. The fact which

In following our formation into the valley which extends from Brunnen to Schwyz, on the eastern shores of the lake of Lucerne, we found that the intermediate succession between the neocomian and the flysch became much more regular and distinct. Situated between the Rigi (that grand accumulation of nagelflue and molasse) on the one side, and the great masses of contorted secondary rocks of Altorf on the other, the valley extending from Brunnen to Sewen and Schwyz is another of these troughs, the sides of which are composed of the secondary limestone, dipping, on both banks of the river, under the nummulitic and shelly deposits. On the northern side the symmetrical order of succession is very clear, as exhibited in this diagram (fig. 11). Commencing the ascending section on the edge of

Fig. 11.

Relations of Cretaceous and Nummulitic rocks at Sewen.

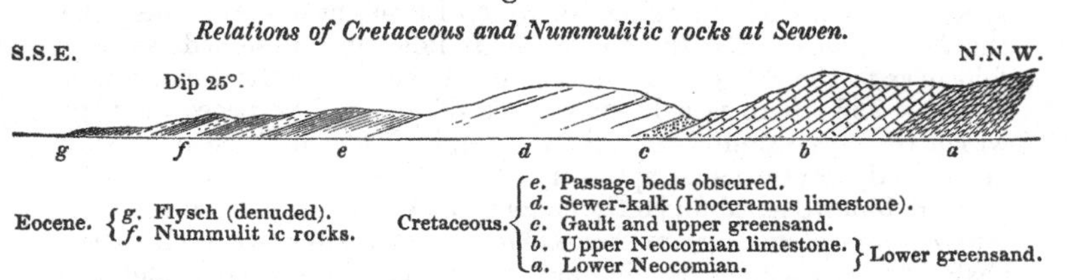

Eocene. { g. Flysch (denuded). f. Nummulit ic rocks.

Cretaceous. { e. Passage beds obscured. d. Sewer-kalk (Inoceramus limestone). c. Gault and upper greensand. b. Upper Neocomian limestone. a. Lower Neocomian. } Lower greensand.

the lake of Lowerz to the north of Sewen, the dark-coloured lower neocomian limestone and shale (*a*) is overlaid by the light-coloured, crystalline, thick-bedded upper neocomian limestone (*b*), in which we detected not only numerous sections of the *Caprotina ammonia*, but also *Hippurites Blumenbachii*, with corals and Echini. Immediately above these is a narrow depression (*c*), in which are softer beds, the equivalents of the gault and upper greensand, with small ammonites and other fossils*.

The next mass which succeeds is the Sewen limestone of the Swiss geologists. This sewer-kalk (*d*) thus resting on upper secondary

is of paramount importance is, that the following species occur in the south of France, the Pyrenees and the Alps, thus identifying the group :—

1. *Nummulites millecaput*, Boubée = N. polygyratus, *Desh.*
2. ———— *planospira*, Boubée = N. assilinoides, *Rüt.*
3. ———— *Biaritzana*, D'Archiac = N. atacica, *Leym.*, N. acuta, *Sow.*, and N. regularis, *Rüt.*
4. ———— *globosa*, Rüt. (var. *Biaritzana*, D'Arch.) = N. obtusa, *Joly et Leym.*
5. ———— *rotularis*, Desh. = N. globulus, *Leym.*
6. ———— *placentula*, Desh. = N. intermedia, *D'Arch.*
7. ———— *lævigata* (Lam.).

1. *Orbitolites discus*, Rüt.
2. ———— *patellaris*, Brünner.
3. ———— *stellaris*, Brünner = Calcarina stellaria, *D'Arch.*

Operculina near to *O. ammonea*, Leym.

* Several fossils of the greensand have been found here by Studer and Escher. The latter geologist was we believe the first to name the overlying limestone Sewerkalk, and to show how a similar limestone occupied a similar place in the canton of Appenzell, and on the lake of Wallenstadt.

greensand, is a whitish grey rock, in which limestone of conchoidal fracture and light colour, forms small and flattened concretions in a light grey earthy base, or rather the shale of this colour wraps in thin coatings over the calcareous undulations, the whole splitting into flagstones six to eight inches thick, and occasionally into much stronger beds. Like the greensand and neocomian limestone on which it rests, this rock seems here to be the exposed portion of a dome, which, as far as can be seen, dips to the N.E., E., S.E., and S.S.E. To the east it is denuded along the bank of the little river, and is extensively quarried as a building-stone. We were fortunate enough to discover (we believe for the first time) Inocerami in this sewer-kalk, fragments of which fossils are to be detected by those who will carefully look for them, even from the lowest beds which rest upon the greensand, to the upper portion of the quarries. The dominant species (two or three specimens of which I brought home) seems to be the *Inoceramus* or *Catillus Cuvieri*. In following these beds as they fold over to the S.S.E., and where they descend into the Muotta-thal at about 25°, there is a hidden space of about fifty paces only, in which the succession is not observed (*e*), but they are then succeeded in perfectly conformable apposition by beds (*f*) of sandy greenish-grained limestone, abundantly charged with nummulites, chiefly the *Nummulina planospira* or *assilinoides*, which alternate with marly shale, which becomes sandier and more flag-like upwards, and are finally surmounted by sandy marlstone charged with Orbitolites, Pectens, &c.

The broad valley watered by the Nieten and the Muotta streams has evidently been excavated in the soft beds of flysch and sandstone superior to the nummulite bands; for after traversing to Ingolboldt, on the external slope of the opposite mountains, the first strata met with at that village, are the very same beds of sandy rotten marlstone with large Orbitolites, Pectens, and casts of other fossils, which there occur in highly inclined strata dipping to the north, and thus form a portion of the opposite side of a trough, as seen in the general section (fig. 12). The flank of the ridges extending from Brunnen up the left bank of the Muotta is much obscured by woods, fallen cliffs, and vegetation; but there are spots in which portions of the nummulitic rock are also seen to be underlaid by the sewer-kalk, greensand, and by upper and lower neocomian, the latter forming the nucleus of the great dome-shaped calcareous mountain Morschach, on the east side of the Altorf branch of the Lake of the four cantons, immediately to the south of Brunnen (fig. 12) *.

Reserving for another part of this memoir the consideration of the enormous flexures and breaks to which this whole series of rocks, together with the jurassic limestones, have been subjected at the upper extremity of the Altorf lake, I would now merely remark, that the woodcuts figs. 11 and 12, the one detailed, the other general, clearly indicate that the Sewen limestone (*d*), with its Inocerami, lying between

* Near the spot called Gumpisch, this lower neocomian dark limestone is loaded with *Gryphæa Couloni*, *Rhynconella* (*Terebratula*) *depressa* (D'Orb.), and *Spatangus retusus*.

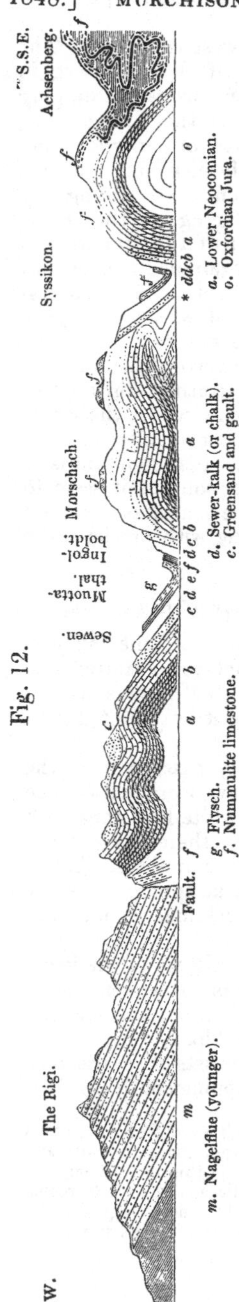

Fig. 12.

the upper greensand (*c*) and the nummulite rocks (*f*), is precisely in the same place as the inoceramus limestone of Thones in Savoy before described, and that both are clearly representatives of the white chalk of Northern Europe.

This sewer-kalk rises up on nearly all sides of the beautiful valley of Schwyz. I refer to it the grand red and white peaks of the Mythen*, which overlook the town of Schwyz, so well known to all lovers of the picturesque. These masses of red and mottled grey and white limestone strongly resemble the scaglia or Italian equivalent of the chalk, and have no sort of resemblance to any other known limestone in the Swiss Alps. They also clearly overlie all the older limestones, jurassic and neocomian; I therefore unhesitatingly refer them to the white chalk; and the more so because they are linked on to the superior nummulitic and flysch formations. On the northern flank of the smaller peak, in ascending to the Hacken pass, we crossed over masses of schist and impure limestone with white veins, which formed the external envelope of the slope, and next over green-grained calciferous grits with *Nummulites planospira, N. rotularis, N. Biaritzana*, and *Orbitolites discus*; the thick shells of the latter resembling little layers of calc-spar; but we also detected a specimen of Inoceramus in the fragments of limestone which had fallen from the cliffs. The upheaval, however, of the Mythen has been accompanied by so much dislocation around it, and such enormous subsidences have taken place on the taluses, that no regularity of succession can be detected; nor could the above order be stated if the adjacent rocks when in their normal positions (as before cited in fig. 11) had not afforded us a true key to the structure of the tract. There is, in fact, just the same appearance of a general inversion of the formations on the

* We passed the Mythen on the north by the Hacken pass in our route to Einsiedeln and returned by Brunnen and the Holzeck pass. The latter is the grandest scene, and is the point from whence the summit is alone accessible.

eastern shore of the lake of Lowerz, as on its western banks. The great accumulations of nagelflue and molasse of the Rossberg in the one case, like those of the Rigi on the other, instead of dipping away from the Alpine centre, plunge towards it; the younger tertiary rocks seeming to be the oldest by their order of superposition; whilst the nummulitic and flysch formation is broken and squeezed up against the cretaceous rocks of the Mythen. In stating my belief that the *summits* of the Mythen are of the age of the chalk, I would not, however, infer that the lower part of the mountain is also of this age; for at the eastern face, when examined from Brunnen, in the valley leading to Einsiedeln, it presents a great succession of underlying massive terraces, the lowest of which are highly altered, siliceous rauchwacke limestone, with partial dolomitic veins not unlike the "cargneule" of Savoy. In the vertical faces of the limestone beneath the red peaks, the lines of stratification are obscurely perceptible, seeming to pass to the S.S.W.; whilst the whole is traversed by highly inclined joints resembling a rude cleavage, the planes of which plunge 70° to the N.N.W. As this mountain is to a great extent inaccessible, and as the lower portions of it seem to have undergone great modification, it is one of the countless examples which the Alps offer, of the difficulty of defining with precision the downward limits of formations.

Nummulite and Flysch Rocks of the Environs of Einsiedeln.

Vast masses of flysch* lie between the Mythen and the valley of Einsiedeln; and to the west and south of that town, terraces of nummulitic limestone rise out from beneath the chief masses of such flysch, and are seen, at the same time, to be strictly united with them.

I have already stated that in Savoy, the only passage from the chalk upwards with which I am acquainted, shows a gradual change in the colour and texture of the rocks from the white limestone with inocerami into the brown sandy nummulitic rock; there being there, as far as I could see, but one band charged with nummulites. Again, in the section near Sewen, as we have seen, there is an apparent passage between the uppermost beds of the inoceramus rock and the nummulitic strata above them.

In the environs of Einsiedeln, however, the thickness of the lower portion of the nummulitic group increases, and in subsequent pages it will be shown how such development becomes still more striking in the canton of Appenzell, and in the Bavarian Alps.

The Schwendberg to the west of Einsiedeln consists of several buttresses of hard brownish nummulitic limestone dipping south (fig. 13).

* I need not repeat the mineral description of the flysch, except where it offers some new features. The generic word applies to the group associated with and overlying the nummulitic rocks, which is chiefly composed of thin-bedded, impure, dark grey limestone with white veins, schists, both argillaceous and calcareous, marls, micaceous sandstones, sometimes green-grained, but more frequently diversified by small black grains, with fucoids and a few casts of shells and fishes' teeth in its lower parts.

In the lower mass, composed of a greenish limestone (f^1), nothing but nummulites are visible, chiefly the *N. millecaput* (Boubée) ; then come greyish blue bands (f^2) with other species of nummulites ; next a considerable thickness of marls, sands, &c. (flysch), surmounted by strong reddish and greyish nummulitic limestones (f^3). In short, from the bottom to the top of the nummulitic portion of the series, there were intercalations of strata having all the characters of flysch. The great overlying mass (*g*), however, has been alone styled such by Swiss geologists, and it here spreads in vast thickness over the adjacent mountains.

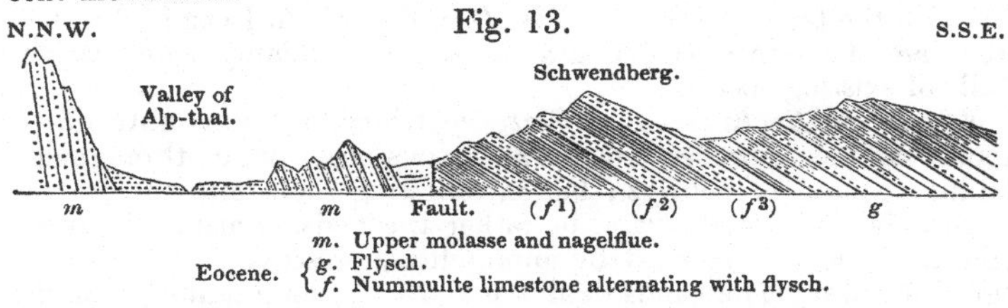

Fig. 13.

m. Upper molasse and nagelflue.

Eocene. { *g*. Flysch.
f. Nummulite limestone alternating with flysch. }

In ascending the little valley of the Sihl from Einsiedeln the same relations of ridges of nummulitic limestones and flysch are still more clearly exposed to the east and west of the village of Gros, where they have also a dip S.S.E. The Sattel mountain on the east side of this valley exposes on its flanks three or four prominent bands of the nummulitic rock, all dipping to the S.S.E., separated from each other as well as overlaid by considerable thicknesses of flysch (*i. e.* of sandstone, limestone, shale, and schist).

The lower nummulitic limestones visible are dark grey, reddish, and greenish grained, in which occur the large echinidæ of Kressenberg, together with ostreæ, small nummulites, and large orbitolites. Then intervenes a great mass of shale and sandstone, followed by a second nummulitic limestone and another zone of flysch, and that again by a third nummulitic limestone. In this last-mentioned mass I was much struck with the strong coincidences between some of the coarser nummulitic limestones and the limestone of the so-called "flysch" of many parts of the Alps. They were, in fact, precisely the same thin-bedded, dark grey, sandy limestones with white veins, and occasionally with so many grains of green earth as to become a green calcareous grit ; the only difference being that the flysch was void of the nummulites and fossils which distinguish the other. These strata, covered by thick beds of grey limestone, pass upwards through shale into fine micaceous flaggy grey sandstone, and thence up into the great series of the so-called "flysch." A similar intercalation and association is indeed quite as instructively seen just above the village of Gros on the west side of the valley.

The chief fossils of the nummulitic bands of this district are the *Nummulites planospira* (Boub.) or *assilinoides* (Rütt.) ; *N. millecaput* or *polygyrata* (Desh.) ; *N. Biaritzana* (D'Arch.) or *regularis* (Rütt.) ; *Operculina*, apparently a large new species ; *Orbitolites discus*, and *O. parmula* ; Pectens, large Ostreæ, and some few univalves, and occasionally the large *Conoclypus conoideus* and other Echinoderms that characterise the deposits of Sonthofen, Kressenberg, and other places in Bavaria and Austria.

All the dips of the rock-masses in this tract are inverted; for the molasse and nagelflue of Einsiedeln being the eastern prolongation of the Rigi and Rossberg equally plunge S.S.E., and seem absolutely to be overlaid by its older neighbour the nummulite limestone and "flysch;" the latter formation being in its turn so thrown over that its younger member lies against or dips under the secondary rocks. In this manner the oldest portion of the nummulitic group is in contact with the tertiary conglomerates (fig. 13), which I shall hereafter prove are the upper part of a great series containing some marine shells of existing species!

Nummulitic Rocks and Fish Slates of Glarus.—Nummulitic limestones reappear in broken troughs at various other points throughout the little canton of Schwyz; but in following them into the canton Glarus, the associated strata, or rather the beds immediately overlying the zone characterized by nummulites, presents a striking zoological feature. The bands of flysch above the nummulites, indeed, as in many other places, contain fishes' scales and teeth, particularly certain dark schists and marls of Savoy and various parts of Switzerland. In Glarus, however, and notably near Engi in the valley of the Sernft, where these black beds have undergone much induration, they are largely quarried under the name of Glarus slates, and are well known to collectors for the numerous fossil fishes they contain. On visiting the quarries I found them totally void of any slaty cleavage; the so-called slates being true calcareous flagstones with a few diagonal veins of white calc spar. They dip away 30° and 40° E.S.E. from the face of the ridge of the most ancient rock of this tract, usually called the Sernft conglomerate. At the spot on the west side of the valley, where the fish beds are quarried, there is no visible relation to any nummulitic rock; but the same calcareous flagstones with white veins, and which are clearly one of the numerous varieties of "flysch," can be followed up the valley of the Sernft to heights of 1000 feet and more above the water-course; and to the east of Elm they are associated with and overlie strong bands of nummulitic limestone. This position was clearly seen by M. Escher and myself as we ascended from Elm to the high pass of Martin's-loch. In treating of some of the contortions, inversions, and breaks of the Alps, I shall have occasion hereafter to return to the consideration of this very remarkable tract; but whether the strata be there overthrown "*en masse*" or not, has nothing to do with the present question; for the calcareous flagstones identical with the fish beds quarried as Glarus slates, and which are in truth a direct prolongation of them, are fairly dovetailed between two courses of nummulite limestone, in the lowest of which I perceived the large *Nummulites millecaput*, and in the other a greenish-grained deep-coloured siliceous limestone with another and smaller nummulite, both of which occur in numerous places in association with all the other fossils of the group and regularly overlying the cretaceous rocks.

These Glarus slates were formerly considered, from their mineral character, to be of high antiquity, and great was the surprise of most geologists when in the work of Agassiz the species of fishes contained in them were classed with so new a formation as the chalk. I now go farther, and assert that, by geological position and association with the nummulitic strata, they are certainly eocene, and possibly not of older date than the lowest portion of the London clay. Nor is there any evidence in the characters of the ichthyolites of Glarus to contravene this inference, but on the contrary much to sustain it. The Palæorhynchum, Acanus, Podocys, &c., are, it is true, extinct genera, but they are also peculiar and unknown in any cretaceous deposits; whilst the Fistularia, Vomer, Osmerus, and Clupæa* have not only never been found in any secondary rock, but are absolutely living genera. Even then, if we had no geological or stratigraphical evidence, one might be fairly led, by the identifications of Agassiz alone, to conclude that a formation including smelts and herrings (there being three species of the latter) was of tertiary age, by the approach of its fauna to the present order of things. The palæontological inference is further sustained by these slates or flagstones containing the bird *Protornis Glariensis*, Herman v. Meyer, and the tortoise *Chelonia Knorrii*, Her. v. Meyer.

In fact there need be no more difficulty in viewing these fish beds of Glarus as tertiary, than the black carbonaceous hard limestones and schists and flysch of the Diableretz.

Nummulite and Flysch Rocks in the Grisons.—M. Studer has shown that large portions of flysch in the Grisons have been converted into a crystalline gneissose rock; but I would now state, that whatever be their irregularities of position in the interior of that canton, and to whatever metamorphisms they may there have been subjected, they unfold themselves with symmetry and regularity in their normal order between the valley of the Rhine and the baths of Pfeffers. In the gorge of the Tamina, to the south of these baths, a clear succession is seen through Oxfordian, Neocomian, and upper cretaceous rocks, which latter pass under nummulite rocks; the baths being situated in vast masses of flysch interlaminated with nummulites, as seen in a section which I made in company with M. Escher. Here again many of the black flags are absolutely identical in mineral characters with the so-called slates of Glarus, and although no entire ichthyolites have been discovered in them, they contain the teeth of fishes.

Sections of the Cretaceous and Nummulitic Systems on the north side of the Lake of Wallenstadt, and in the Hoher Sentis of Appenzell.—Whilst the section of the Tamina and the baths of Pfeffers show the ascending order from the cretaceous rocks up into the nummulitic limestone and flysch, by far the largest and clearest exhibitions of the whole succession in Switzerland are displayed on the plateaus on the north bank of the lake of Wallenstadt†, and in the environs of the

* See Agassiz, Poissons Fossiles, General Table, tom. i. p. xxxiii, where forty-two species of fishes are named.

† When in the Tyrol with M. von Buch in the previous autumn, he assured me

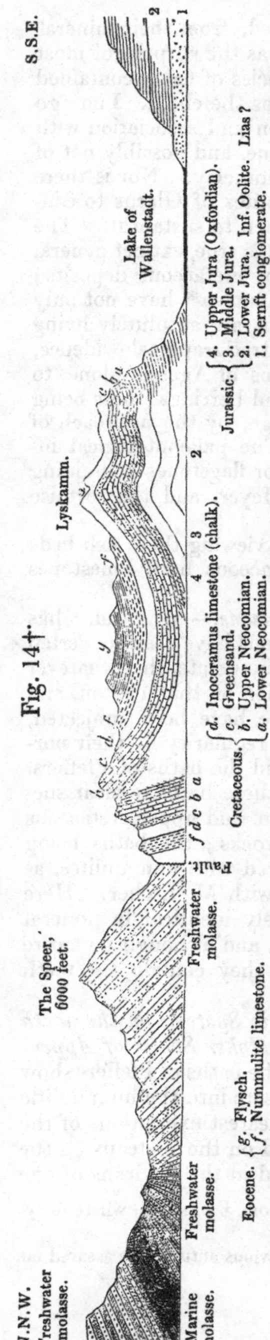

Fig. 14†

still loftier Hoher Sentis and the high tracts of Appenzell; a district which is rendered classical in geology by the recent labours of M. Arnold Escher de Linth. We have there not only that series which has been already so much dwelt upon, from a low horizon in the Jura limestones through the neocomian to the cretaceous series including the inoceramus limestone, but also a very complete exhibition of nummulite rock and flysch. Seeing that M. Escher had so fully made himself master of all the dislocations as well as all the regular successions both of the plateau of Wildhaus and of the Hoher Sentis, I could only pretend to offer one, I trust not unimportant addition to his valuable contributions, by bringing to his notice a band between the inoceramus and nummulite limestones, which I consider to be of value in demonstrating a lithological transition from the cretaceous system, properly so defined, to that which overlies it. I also urged him to adopt my method of classifying the nummulitic and flysch rocks as lower tertiary, and no longer to include them in the cretaceous series*.

In travelling from Mells near Sargans to Wallenstadt and Wesen, a clear ascending order of the strata is visible, from the "Sernft" conglomerate (the most ancient rock of the region) at Mells (fig. 14). This by reference to his well-filled note-book, that this would be found the grandest and clearest of all the Swiss sections in explaining the true overlying relations of the flysch and nummulites to the cretaceous rocks.

* In the tabular view of his memoir entitled 'Gebirgskunde,' appended to the description of the canton Glarus by Professor Heer, M. Escher gives the following names in ascending order of all the strata of the "Kreide-bildungen":—1. Spatangus kalk (Studer) or Lower Neocomian. 2. Schratten kalk or Upper Neocomian. 3. Turriliten-Etage or Gault. 4. Sewerkalk or chalk. 5. Nummuliten Etage. 6. Flysch and Dachschiefer von Plattenberg.

† Although this diagram, from the pencil of M. Escher, exhibits a thin course of it only, the zone of gault or upper greensand exists in full force around the Sentis. In the valley of the Rhine, and extending to Eichberg, it appears in insulated hummocks charged with Turrilites, *Inoceramus sulcatus* and other fossils, which with many forms gathered from the different overlying rocks around the Hoher Sentis, Prof. Brünner and myself examined in the rich museum of the Rev. Mr. Rechsteiner of Eichberg.

conglomerate (1), which is in parts a purple and green spotted glossy schist, in parts a millstone grit, passing into a conglomerate with pebbles of white quartz in a talcose base, dips to the north and passes under the limestones which form the great escarpment of the chain called "Kurfürsten." In that escarpment the Lower Jura and Oxfordian formations (2, 3, 4) are surmounted by the three members before cited of the cretaceous system, viz. neocomian with its two divisions (*a* and *b*), gault or upper greensand (*c*) and inoceramus limestone (sewer-kalk) (*d*). The same overlying succession is seen on the northern shore of the lake of Wallenstadt, the mountains in which are a western prolongation of the Kurfürsten. Above all this and to the north is the upland depression or trough of Wildhaus, in which the inoceramus limestone (*d*) is covered by nummulite limestone and flysch (*f* and *g*). The latter deposits rise up to the north of Wildhaus in a basin-shape with a reversed or southern dip, and then equally repose on inoceramus limestone, which is succeeded by the gault and neocomian limestone (*a* and *b*), the latter culminating in the chief summits of the Hoher Sentis. That mountain group, the highest points of which are near 8000 English feet above the sea, and which forms by far the most remarkable promontory along the whole outer zone of the Alps, is highly instructive in the full development of all the cretaceous rocks from the lower neocomian to the inoceramus limestone, as seen in the Alte Mann as well as in the other chief summits.

I shall hereafter advert to its escarpments when speaking of the great flexures and fractures of the chain, where the tertiary nagelflue is apparently brought under the masses of secondary limestone. I will now briefly state, that on the north flank of the Kamor, a north-eastern promontory of this group, and again on the Fähnern mountain beyond it, there are natural sections which exhibit the supracretaceous succession (see fig. 15). The last boss of the sewer-kalk of the

Fig. 15.

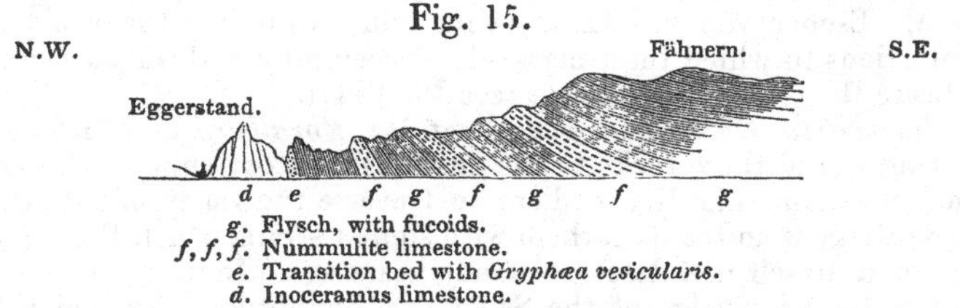

g. Flysch, with fucoids.
f, f, f. Nummulite limestone.
e. Transition bed with *Gryphæa vesicularis*.
d. Inoceramus limestone.

Hoher Sentis, prolonged in a low ridge to the N.E. of Weissbad, constitutes a scarp immediately overhanging the little hamlet of Eggerstand, where, in the form of thin-bedded white scaglia, it plunges rapidly to the S.E., and is immediately covered by slightly micaceous shale and bluish grey impure limestone with white veins. This rock, which already resembles a variety of "flysch," passes up into a sort of sandy marlstone with some green grains associated with a dark indigo-coloured schist, in which occurs the same species of Gryphæa, *G. vesicularis*, to which I shall subsequently call attention in describing the sections at Sonthofen in Bavaria, where it occupies a like place.

The next bed (*d*) is a green calc-grit charged with nummulites and orbitolites. So far all is clearly seen on the sides of the broken ravines descending near Eggerstand. In ascending the Fähnern, or rather in coasting its western face obliquely towards its summit from the ravines above-mentioned, you next pass over a considerable thickness of schists and sandstone or flysch, and then reach another and the chief band of nummulite limestone which ranges along to Schwarzen Eck. This is a very green-grained, sandy limestone, which when bruised by the hammer is rendered grass-green, and contains *Nummulites globulus, N. globosa* and *N. millecaput,* Boubée, together with Orbitolites and several forms of Pecten and the usual fossils of the group.

The inclination of the strata gradually decreasing as the axis of disturbance is receded from, the nummulite bands graduate upwards into other beds of flysch in which no animal forms are visible, and finally towards the summit of the hill into finely laminated, light-coloured calcareous flagstone, on the laminæ of which are numerous impressions of fucoids of at least three species, viz. *F. Targioni, F. intricatus,* and a new species with broad fronds, described by Prof. Brünner as *F. Helveticus**.

In relation to these fucoids, I may here observe once for all, that throughout the Savoy and Swiss Alps, and indeed I now believe generally all along the *northern* face of the chain, they occur in a zone superior to the chief masses of nummulitic limestone. The beds in which they occur are, however, so linked on to the inferior members of the group in numerous natural sections, (there being no instances of dislocations or unconformity between the one and the other with which I am acquainted, except on lines of fault,) that I necessarily consider them to form one natural group with the nummulitic rocks on which they repose. In treating of the flexures and breaks of the calcareous mountains of the Alps, I will hereafter produce a series of transverse sections across the group of the Hoher Sentis, as prepared by M. Escher von der Linth, which in exhibiting the wonderful contortions to which these masses have been subjected will also clearly indicate the order of the strata (see Pl. VII.).

Nummulitic Rocks and Flysch of the Voralberg and Allgau.—Having traced these rocks to the north-eastern extremity of Switzerland, it became highly expedient to traverse the valley of the Rhine above Bregenz and connect them with similar strata, which Prof. Sedgwick and myself had described many years ago. In fact, I could not acquire the knowledge of the Savoy and Swiss succession which has now been detailed, without seeing clearly that our former classification of the nummulitic rocks and flysch of Dörnbirn in the Voralberg, and of Sonthofen in Bavaria, and of various places in Austria, with the cretaceous system and greensand, must be changed.

The nummulite beds near Dörnbirn on the right bank of the Rhine have here been correctly described as apparently dipping southwards

* Professor Brünner has also shown that the *Fucus Brianteus* (Villa) of the Briançon on the flanks of the Milanese Alps is identical with a species found in the Gurnigel sandstone or flysch near Berne.

under the great calcareous masses of the Stauffen*. On visiting this spot with Prof. Brünner I found dark greyish, white-veined limestone with schist or shale in the mass now visible *in situ*, which, if the fossils were omitted, would be "flysch," surmounted by other bands of schist or shale and sandy green-grained limestone passing into a grey rock, and again shale and schist with thin stone bands of "flysch." In the lower limestone were small-ribbed Pectens, large Ostreæ, Terebratulæ, Echini, and many Nummulites. The higher portion of the upper band is characterized by *Orbitolites discus* and the *Nummulites globosa*. In the lower mass is the highly ferriferous band formerly worked for iron, which is a perfect congeries of the *Nummulina planospira* or *assilinoides* and *N. placentula* (Desh.). These fossils are precisely those of the Fähnern mountain on the opposite bank of the Rhine; whilst in the association of iron with the nummulites it is clear that it is the direct western extension of the still more ferruginous zone of Sonthofen in Bavaria.

These nummulitic and flysch beds apparently dip under the secondary limestone. The mural escarpment of the Breitenberg, a counterfort of the Stauffen, which seems to be the upcast mass, consists chiefly of neocomian limestone, and in the part to which we ascended with some difficulty through the thick woods we found the *Spatangus retusus* of the *lower* member of that formation. It is probable that there is really an overlap at this junction as represented in this woodcut, fig. 16, and the point will be discussed in the sequel.

Fig. 16.

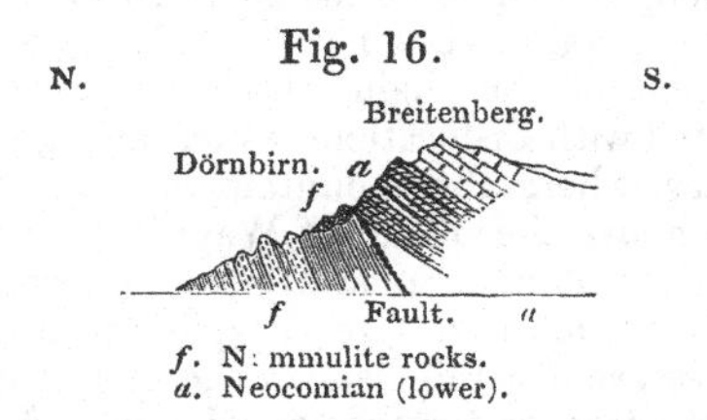

f. Nummulite rocks.
a. Neocomian (lower).

Sonthofen Iron Mines, and the Grünten Mountain in Bavaria.—The symmetrical order of succession so clearly exposed on the outer flank of the Fähnern and at other points around the Hoher Sentis, and which is obliterated along the great line of fault at Dörnbirn near Bregenz, is strikingly and instructively resumed in the Grünten mountain, situated between Immenstadt and Sonthofen in Bavaria.

Sections of this mountain were published in the communication so often alluded to†, but they were defective in not presenting any well-defined geological horizon either in the inferior or superior strata. It is true that Prof. Sedgwick and myself then discovered greensands with unquestionable British cretaceous fossils, and we stated that these were surmounted by the scaglia or equivalent of the chalk. But the transition downwards from that which really represents the gault and upper greensand into the fossiliferous limestones, now

* Trans. Geol. Soc. Lond. vol. iii., 2nd series, p. 325, and pl. 36. fig. 3.
† *Ibid.* pl. 36. fig. 4.

known to be neocomian or lower greensand, was wholly omitted; for, as before said, the neocomian fossils were then unknown, and these rocks were considered to be of jurassic age. On the other hand, the transition upwards from the equivalent of the chalk into the nummulitic grit, and thence into the flysch as an overlying mass, was imperfectly explained. In short, having returned to Sonthofen and the Grünten after an interval of eighteen years, and immediately after I had made a consecutive series of sections in strata of this age throughout the Savoy and Swiss Alps, I looked at the masses with a different eye to that with which I viewed them when the only Alpine bases known to me were the rock-masses (often inverted) on the north flank of the Austrian Alps. Even formerly, however, when treating of the flysch with fucoids of this valley of Sonthofen, Prof. Sedgwick and myself offered our sketch as a provisional arrangement only; stating that a more minute acquaintance with the fossil history of the Alps might hereafter lead geologists to a better-defined subdivision of these groups. Profiting, therefore, by the increase of this very fossil knowledge and by a study of the best types in other parts of the chain, and correcting my former views, I now offer sections which I consider to be as clear, copious and instructive, in explaining the succession from the cretaceous to the nummulitic rocks, as any with which I am acquainted.

The peaked and remarkable calcareous mountain called the Grünten (5923 French feet high), which stands out boldly between Immenstadt and Sonthofen, and there forms the eastern side of the valley of the Iller, has a general direction from N.E. to S.W. This direction, oblique to that of the chain which trends from W.N.W. to E.N.E., is connected with dislocations which affect all this tract. On the north-west face, where the mountain is washed by the Iller, it throws out a spur above the village of Wagneritz or towards Immenstadt; to the north it abuts against a mass of tertiary molasse; on the south-east it is divided into several jagged peaks, the precipitous walls of which preserve a parallelism to the main ridge of summits; whilst on the south-west, or towards Sonthofen and the upper valley of the Iller, round-shaped buttresses diminishing in height expose an excentric arrangement of strata in ascending order. In a word, the general escarpment of the mountain is to the north-west and north-east, and the prevailing dips of the strata to the south-east and south-west. The best general section may be described as that which exposes an ascending order from the elevated escarpment near Rettenberg on the north-east, to the plain of Sonthofen beyond the village of Burgberg on the south-west. As in proceeding upon this ascending section the strata towards the south-west are found to mantle round and overlap the chief nucleus, it follows that lines drawn either to the south or west of the sectional line will exhibit similar successions. Thus, on the south-eastern face of the Grünten, vertical walls of jagged limestone, which diminish in height from the summits of the mountain to the valley of the Starzlach, expose precisely the same ascending order of strata as that which is seen in the masses that fold over at less high angles towards Burgberg and Sonthofen.

A glance at this diagram, fig. 17, will sufficiently explain the case.

Fig. 17.

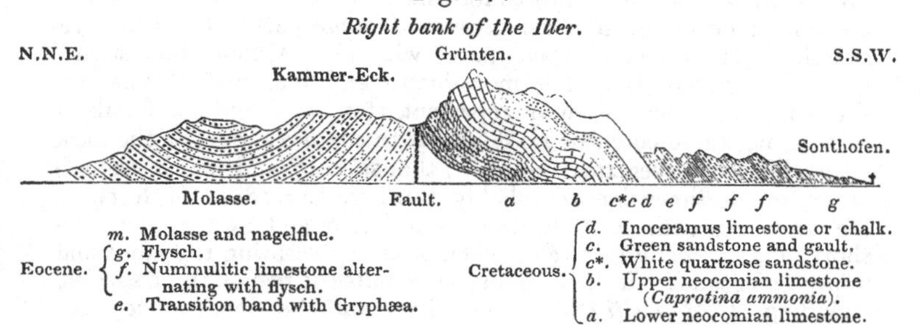

The lowest visible rocks, as seen in the escarpments on the north and north-east faces of the Grünten (*a* of fig. 17), are shaly, dark grey, thin-bedded compact limestone, with a little iron and nodules of black flint, alternating repeatedly with dark shale. Some of the beds contain so much chlorite, that, like rocks in two other zones higher in the series, they become grass-green when bruised by the hammer, though previously they are simply dull grey calcareous grits or impure limestones with schists. With the exception of an ammonite, M. Brunner† and myself found no fossils in this rock. There can, however, be no doubt that it is the lower neocomian of Swiss geologists, which lithologically it resembles, and like which it graduates up into, and is at once overlaid by, the true upper neocomian, white limestone. The latter rock (*b*), which, as has been stated, forms so clear a horizon throughout large regions of the external calcareous chain of the Alps, is here, as elsewhere, a thick-bedded, compact, light grey limestone, weathering white in the cliffs; the surface being distinguished by innumerable white lines, occasionally defining the segments of the shell of the *Caprotina ammonia* and other fossils. Usually, indeed, these fossil outlines are the hardest portions of the rock, and stand out in the form of chert. Veins of white calc spar also traverse the strata. This white limestone or upper neocomian constitutes the highest point of the double-peaked Grünten, a narrow broken wall of limestone trending from north-east to south-west, the beds of which dip rapidly to the south-west. The consequence is, that in following the top of the crest from these limestone summits to another point called the Hohe Wand, where a cross is erected, and thence down to the highest houses in an upland gorge, called the Gundalpe Hütte, you pass successively from the neocomian above described to other overlying formations. The rock (*c*) immediately resting upon the upper neocomian limestone is a lightish grey, brownish, and even a whitish siliceous or quartzose sandstone. Finding this rock in other sections on the sides of this mountain, as well as upon the summit, and always in this position, viz. overlying the limestone recognised as the upper neocomian of the Alps; and,

† This name has been misprinted Brünner in the preceding pages.

further, seeing that in all situations it is capped by a zone of dark green, schistose sandstone which contains fossils of the gault or upper greensand, I was induced to think that it might represent the upper portion of our English lower greensand, some parts of which it resembles. It may also be compared with the "Quader Sandstein" of Saxony, except that it is more brittle and quartzose. Whatever the sandstone rock (*c*) may represent (for we found no fossils or casts in it), there could be no doubt as to the next zone, or the dark shale and deep green sandstone (*c**) that succeed, and which, though of no great dimensions (probably nowhere exceeding 50 feet), is the same excellent fossil horizon as in Savoy and Switzerland. In short, it is the band so often spoken of as representing the gault and upper greensand. In it we found ammonites of two or three species, including *A. Mantelli* (Sow.), Turrilites, and the small *Inoceramus concentricus* (Sow.).

Some of these fossils also occur in a lateral spur of the Grünten, towards the village of Wangeritz, and others on the external face of the great dome-shaped mass which, in the ravines to the east of Burgberg, exhibit this dark green sandstone passing up into a thin band of hard, compact, cream-coloured limestone impregnated with chlorite; in short a hard "craie chloritée." The green sandstone is extensively quarried on one of the shoulders of the Grünten to the north side of the great depression called the Vust†, between the mountain and the nummulite ledges (*f*) that run down to Burgberg, and when worked out is really a very striking band. It is a mottled rock, and frequently owes this appearance to branching flattened stems, which may be Alcyonia.

The inoceramus limestone (*d*), with its chloritic base, above alluded to, forms a wrapper of great thickness over the green sandstone or gault, and constitutes the external coat of the mountain on its western and south-western faces. It is largely and clearly exposed in the breaks on the sides of the upland depression of the Gundalpe Hütte, above the Vust ravine, from whence it rises up to the summit called the Hohe Wand, the cross of which stands on it, and very near its junction with the inferior zone of green sandstone. In parts it is of the colour of the sewer-kalk, *i. e.* a light grey or green colour; but above the Gundalpen, or between these châlets and the Hohe Wand, it graduates into limestone as red as the scaglia of Italy, or of the Mythen mountain near Schwyz. Throughout its matrix are numerous fragments, occasionally almost entire shells, of large thick-shelled inocerami. This rock of the Grünten, so clearly in the posi-

† Many of the fossils, so called, of Sonthofen, collected by the Bergmeister and other persons, have been found in the beds of this broad torrent called the Vust. Now, as the waters which flow into it traverse all the strata in the cretaceous succession, and these flank the nummulitic beds, geologists will readily understand how Prof. Sedgwick and myself were formerly led to believe, by the inspection of such collections, that nummulites occurred in the same beds with ammonites and belemnites and small inocerami, the *green sandstones above and below the equivalent of the chalk often closely resembling each other*. I have now satisfied myself that here, as elsewhere throughout the Alps, nummulites are unknown below the surface of the inoceramus limestone.

tion of the chalk of North Europe and of the scaglia of North Italy, is of very considerable thickness, certainly several hundred feet.

The largest superficies in which the inoceramus limestone is exposed, is around the dome-shaped masses, the external faces of which dip rapidly down into the great ravines north-east and east of Burgberg. In the latter we perceived it to be overspread by a thin course of dark greyish, fatty marl, in which we detected one inoceramus. This band is immediately surmounted by marly and incoherent, slightly micaceous, thin-bedded sandy shale, which here has been largely denuded, and above Burgberg is exposed in a transverse depression between the Grünten mountain on the one hand and the lower nummulitic ridges on the other.

This hollow space (the Vust) between the external face of every stratum to which the terms 'chalk' or 'cretaceous' can rigorously be applied, and the lowest band of nummulitic limestone, is occupied in its lower portion by the small micaceous shale and schist before mentioned, which is succeeded by a greenish sandstone associated with impure greyish limestone and dark grey shale. These beds, particularly the sandy impure limestone, contain the same *Gryphæa vesicularis* which has been remarked as lying between the inoceramus limestone and the nummulitic rocks of the Fähnern in Appenzell. Here, however, this intermediate band of green sandstone, schist and limestone (*e* of the diagram) is vastly more expanded. If the section be made in the regular ascending order of the mountain (fig. 17), as followed from its main escarpment, over its summits, down the Gundalpe, and across the Vust to the nummulitic ridges east of Burgberg, this intermediate group (*e*) is seen to be perfectly conformable to the inoceramus limestone beneath it, and to the lowest nummulitic rock above it. Equally is it conformable if another section at right angles to the above be made from the Grünten to the valley of the Starzlach, a little to the south of the chief mines, and where a rivulet descends from the mountain (see fig. 18). In

Fig. 18.

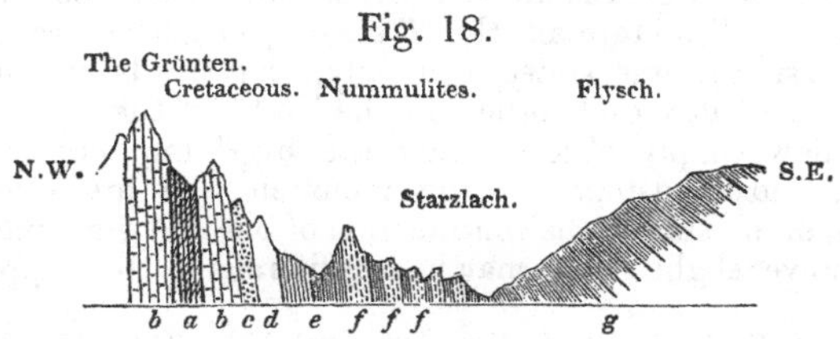

this section the beds are more nearly vertical, and necessarily occupy very small horizontal spaces. The same order being followed from centre to flank, *i. e.* from the neocomian through the greensand and cretaceous strata, the explorer does not fail to observe a great thickness of bluish grey, slightly micaceous marls, and marlstone associated with a sort of greensand, and beds of impure grey limestone with white veins (*e*), in which we again detected the same Gryphæa as in similar strata in a like position in the other section near Burgberg, p. 205.

The Gryphæa to which I have now so much alluded, is considered by Mr. Morris, M. D'Archiac, and all the conchologists who have examined and compared it since my return to England, to be the *G. vesicularis*, a fossil of the upper chalk of England, and which in the south of France is common to the white chalk and the lowest nummulitic zone. It was either this species or its representative Gryphite, which Professor Sedgwick and myself collected at Matsee, north of Salzburg in Austria, where it occurs in strata similar to those of the Grünten and Fähnern mountains, and where it is equally surmounted by nummulitic limestones with large Echini and Pectens*. If then we are guided by fossils, we ought to group this band or intermediate bed (*e*) with the cretaceous system, although its beds have already assumed to a great extent the lithological characters of the overlying nummulitic greensands and flysch into which they make an imperceptible transition. In the Fähnern mountain, indeed, the same Gryphæa continues to pervade the ascending strata until it is associated with nummulites; whilst in the Vicentine, another species of Gryphæa approaching to the *G. columba*, mounts, as is well known, into strata in which not only nummulites, but many true eocene shells occur. These Gryphites (perhaps two or more species) characterize, therefore, the zone of transition between the secondary and tertiary rocks of the Northern Alps.

Nummulite Rocks and "Flysch."—After a clear exhibition on the sides of the torrents of several courses of the above-mentioned strata with Gryphites, some of which lithologically resemble the flysch above the nummulites, these beds (*e*, fig. 18), dipping 60° to 70° south-east, are seen to graduate conformably into another and somewhat thicker band of limestone of deep ferruginous colour, which is loaded with myriads of nummulites, grains of chlorite being abundantly disseminated in it (*f*). This is the lowest of the several well-known zones of nummulitic iron ore of Sonthofen, and it is charged with the large *Nummulina millecaput* and *N. planospira*, the smaller *N. globosa*, the large Echini with Crustacea (*Cancer Sonthofensis*), Pectens, some Terebratulæ, the *Trochus giganteus*, etc†. On a former occasion it was stated generally, that bands of nummulitic limestone succeeded each other on the banks of the Starzlach, and I would now simply observe, that the overlying schists, impure limestone, and sandstones of the mountain (*g*) are referable to the flysch, or are simply the continuation of one and the same series of strata, however slightly they may be fossiliferous in their upper parts.

* See Trans. Geol. Soc. Lond., New Series, vol. iii. p. 349. The Gryphæa of Matsee is named *G. expansa* by Mr. J. Sowerby. Unluckily the true equivalents of the chalk or limestone with Inocerami are not visible near Matsee. It is now however my belief (though I have not re-examined the country) that all the extensive mass of flysch or Vienna sandstone lying between Matsee, Mondsee, and the walls of secondary limestone on the south, is of lower tertiary age.

† Trans. Geol. Soc., New Series, vol. iii. p. 332. The genera are Pecten, Terebratula, Spondylus, Plicatula, Astarte, Anomia, Isocardia, with large Serpulæ, the well-known large Echini, and the *Cancer Sonthofensis*. It is in fact the same group, most of the species being undescribed, as that which occurs all through the Swiss Alps.

Returning to the chief section (fig. 17), I specially call attention to the ascending succession, as seen to the south-east of Burgberg, above the intermediate gryphite zone (*e*). As on the east flank of the Grünten (fig. 18), so here we see schist and thin bands of dark flysch with white veins, intercalated between the greensand and impure limestone with the Gryphæa and the lower zone of nummulites. The mineralogical transition is here equally perfect, nummulites are also abundant, together with Pectens, Spondyli, and other fossils of the zone, the beds being also ferriferous, but offering some local peculiarities, such as small cavities in the green sandy calciferous grits. This band (*f*) is overlaid by glossy light grey and dark schists, that have been worn into a small depression, which is followed by a second ridge of nummulitic rock. The mass of this is a greenish, yellowish sandstone, or sandy calc grit, which graduates into a hard siliceous limestone containing large Echini, Pectens, Terebratulæ, as well as Nummulites, and is very peculiar from the small flakes of chlorite which occupy the structural divisions of some of the foraminifera. Shale and thin stone bands recur in another slight depression, followed by another course of nummulitic limestone of grey colour, but also containing iron, whereon an ancient castle stands; then another depression in shale, &c.; and lastly a great band of nummulitic limestone of about 150 feet in thickness, which, being thin-bedded, sandy, and subconcretionary in its lower parts, passes up into very thick beds of hard grey limestone charged with *Nummulina millecaput*, Orbitolites, &c. This limestone, when followed to the Starzlach, plunges under other courses of schist and sandy shale, forming part of the great overlying masses that occupy both banks of the river Iller, between Sonthofen and Ober Maiselstein, but which are denuded in the plain of Sonthofen.

I may complete this ascending section of the formations in the valley of the Iller by stating, that although a consecutive ascending order is observable in the hills to the east, the same order cannot be followed without breaks, curvatures and reversals in the chief depression or on its western side. It is manifest, however, that all the sandstones, schists and bastard limestones, which constitute the flysch on both sides of the valley between Sonthofen on the north and the Schwarzenberg, are parts of that great group the lower portion of which inosculates with the nummulitic limestones. (See fig. 19.)

Fig. 19.

Left bank of the Iller, above Sonthofen.

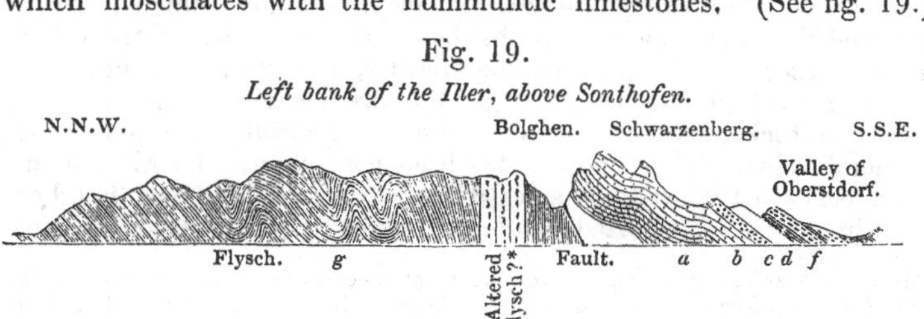

* The small *Nummulina placentula* (Desh.), *N. intermedia* (D'Arch.), in this band, is, I believe, the same species known in the nummulitic limestones of

This upper and much larger division of the supracretaceous formation, which so rarely exhibits fossils, is chiefly characterized by its fucoids, viz. *Fucoides Targioni* and *F. intricatus*. A little to the north of the turnpike and bridge over the Iller, west of Sonthofen, this "flysch" is in the condition of a light-coloured, greenish grey, micaceous sandstone with black grains, in beds from two to four feet thick, and undistinguishable from strata which I shall hereafter dwell upon as the "macigno" of the Italians; one bed of it, an excellent building-stone, being twelve feet thick, and much resembling the "pietra forte" of the Florentines. Intercalated with some of this "macigno alpin," I detected a thin course of nummulitic limestone, the uppermost limit of the nummulites in this region; for in the still higher masses of "flysch," extending by the Bolghen to the foot of the Schwarzenberg near Ober Maiselstein, no traces of other fossils, except fucoids, have been seen.

It is unnecessary to say more on the mineral characters of the overlying group of sandstones, limestones, calcareous grits, argillaceous schist, and calcareous shale and flagstone, which compose the flysch; and after all the details given, I need scarcely remind my readers, that everywhere in the Swiss and Bavarian Alps, where the order has been preserved, this group passes downwards into, and inosculates with, the nummulite limestones above the inoceramus limestone or chalk. When combined with the nummulitic strata (from which I hold them to be inseparable) they constitute therefore one of the grandest formations of the Alps, often rivaling in thickness the whole jurassic limestones, and being of as great thickness as the cretaceous rocks on which they rest.

Altered Rocks of the Bolghen.

Before I take leave of the valley of Sonthofen, I must explain my present views respecting the phænomena in the Bolghen mountain near Ober Maiselstein, where large masses of crystalline rock (having the character of mica schist, gneiss? &c.) were described* as penetrating the green sandstone, fucoid shales and millstone grits of the flysch series. Judging chiefly from Scottish analogies, I formerly thought that these crystalline rocks, which I then believed to be of primary age, had been partially protruded in wedge-shaped and conical masses through overlying sandstones and schists; and I deemed this view the more probable, as on both sides of the valley of the Iller in this part of the district, the strata are not only much convulsed and set on edge, but are partially penetrated by eruptive rocks, and on the east side of the valley contain many mineral veins. That opinion has been controverted by M. Studer, who believes that those masses of crystalline rock in the Bolghen are in truth transported *boulders*, which were included in the

Mosciano near Florence. Besides the prevalent species of nummulites, viz. *N. planospira*, *N. millecaput*, *N. Biaritzana*, &c., Professor Brunner thought he discovered a new species, which he proposed to name *N. Murchisoni*.

* See Trans. Geol. Soc. Lond. vol. iii. p. 334.

"flysch" during the period of its formation. As he mainly supports that view by the example of certain granitic blocks of the valley of Habkheren near Interlacken, and as the interstratification of such boulders or blocks in strata of that age must be very novel to English geologists, I crave permission to digress from the chief objects of this memoir, in order to discuss a point, which, according to M. Studer, is closely related to the structure of the "flysch."

In the valley of Habkheren, on the north bank of the lake of Thun (as in many interior valleys of the calcareous chains of the Swiss Alps), the flysch is squeezed up in a narrow trough with broken and highly inclined strata, portions of which are exhibited on the right-hand side of the hill road which ascends from Interlacken to Habkheren. That these beds belong to the true supracretaceous flysch is undoubted, because in rising up they overlap the nummulitic limestone at the head of the valley, which rock in its turn surmounts the neocomian limestones of the adjacent chain. These "flysch" rocks, in parts pebbly and gritty, in parts schistose, together with the usual shale and thin-bedded, dark, impure white-veined limestones of this series, are there seen to contain truly intercalated geodes and bands of a granitoid character, which re-occur at intervals in a distance of about 150 paces, much in the manner represented in the diagram (fig. 20), the granitic geodes often imitating

Fig. 20.

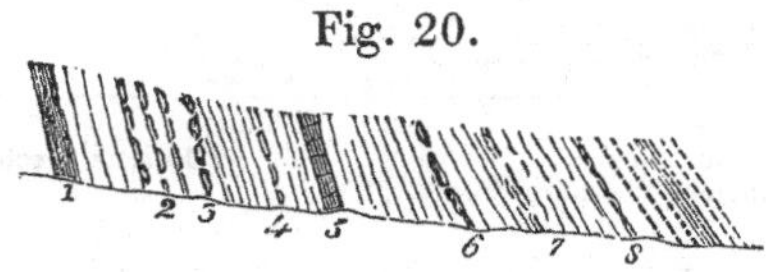

1. Greenish crystalline granitic course.
2. Alternations of schist and impure limestone or "flysch."
3. Granitic geodes.
4. Schist and limestone, &c.
5. Quartzose granitic band.
6. Black schists with calcareous concretions.
7. Schists with granitic concretions.
8. Schists and limestones overlaid by granitoid conglomerates, &c.

in form the calcareous nodules! My attention was first directed to this section by Professor Studer, in company with whom and M. Merian and M. Favre, I visited it. It appeared to me, that the granitoid-like concretions are there intercalated with calcareous nodules, as well as that the thin granitic courses alternate with the schists and impure limestones. The largest of the concretions visible in the course (3) is an oblate spheroid about four feet long by three feet wide; the external zone being more schistose, the interior passing from a paste with large crystals of felspar to a more compact nucleus, one extremity of which seemed almost as if made up of small granite pebbles. The band (5), on the other hand, appeared to be an uniform greenish-coloured granite or granitic gneiss.

I confess that I could not account for such appearances, except by supposing that the granitic matter was evolved contemporaneously with the formation of the sedimentary sandstones and schists which envelope it; the concretionary forms of some of these masses seeming to favour the hypothesis. But whether produced in the same manner as the so-called volcanic or plutonic grits of other regions, by contemporaneous segregation of the igneously-formed particles in the bottom of a turbid sea, or by subsequent partial alteration of the strata through the action of heat and gases, or by transport from other rocks, it is clear that these small developments of granitic matter are contemporaneous with the flysch.

Now, it happens that in the same valley of Habkheren several large granitic blocks also exist, which lying upon the surface of the ancient alluvium, or having been washed into water-courses, have at a distance all the aspect of the usual Alpine erratic blocks, about the transport of which there has been so much discussion*. The largest of these lies on the surface of a boggy meadow, under which is a great thickness of the coarse ancient alluvia on the east side of the rivulet of Habkheren, as explained in the woodcut (fig. 21). This

Fig. 21.

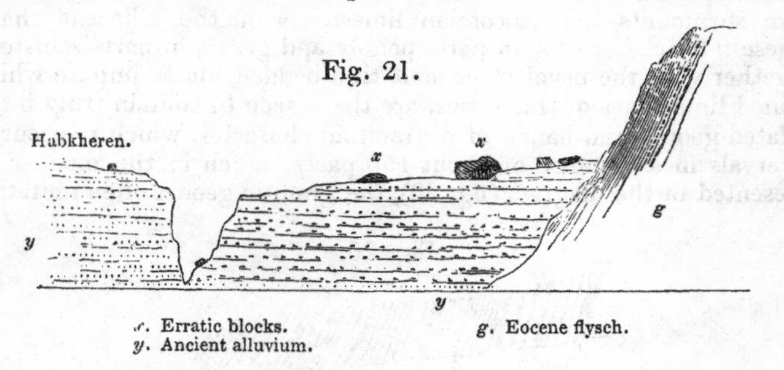

x. Erratic blocks. g. Eocene flysch.
y. Ancient alluvium.

block, so superposed to the ancient alluvia, is about 105 feet long by 90 feet broad and 45 feet high (above the marshy meadow), and has therefore a mass of not less than 400,000 cubic feet. As it consists of a peculiar granite†, now unknown to mineralogists in any part of the Alps, Professor Studer believes that, like the very small geodes and courses alluded to, this block was also included in the formation of the flysch, and that during the disintegration of that rock on the vertical sides of the valley, it has rolled down into its present position.

Extending this view, M. Studer accounts in a similar manner for what he calls the blocks of the Bolghen, *i. e.* that they were derived from pre-existing rocks, and were originally encased in the flysch during its formation. After examining both spots, I cannot adopt this opinion, nor can I regard the great block of Habkheren in any

* In this memoir I shall not enter upon the question of the Alpine erratics, it being my intention, at a future day, to give my opinions concerning their transport and their relation to former glaciers.

† According to Studer, this granite is composed of bluish-white and pink felspar, the latter possibly albite, with white quartz, occasionally weathering yellowish, and dark bronze-coloured mica in small crystals.

other light than that of a huge superficial erratic derived from some parent rock, which has either since been lost by subsidence and buried beneath other deposits, or is hidden from sight under those coverings of snow and ice, which necessarily impede observation over so very large an area of the higher Alps. I fully admit that the small bands of granitic rock above adverted to, are fairly intercalated in the flysch, but the presence of geodes, the largest of which is not above four feet long and a foot wide, can never satisfy me that a monstrous block, containing 400,000 cubic feet, was similarly formed; that block not having the slightest appearance of having ever been a geode. Again, no conglomerates known in any part of the flysch of the Alps exhibit pebbles of more than a foot or two in diameter. But, supposing this block to have been part of a conglomerate, and that it was transported from a ridge of crystalline rock into the flysch during the formation of that deposit, by what agency must we suppose it to have been moved? Certainly not either by solid or floating ice; for the period of the nummulites and flysch was anything but one of glacier action, and was in fact one of considerable warmth.

Seeing then no satisfactory explanation of the deposit of a block of this magnitude in finely laminated sandstone and schist (such as constitutes the flysch of the sides of the valley of Habkheren) (*g* of fig. 21), I necessarily reject the application of such reasoning to the Bolghen. On re-examining that locality (see fig. 19, p. 209) I perceived that the rocks which I had described as millstone grits, greensands and schists, have each of them a persistent strike. Thus, quartz grits, passing into highly indurated schists, the former assuming the vitrified aspect of certain quartz rocks, trend from the slopes above Ober Maiselstein to the summits of the Bolghen on the west. They are, in fact, either vertical or dip 70° to 80° north and south. Now, associated with these, and having indeed quartz rocks on both sides of it, the chief boss of mica schist rock protrudes itself. From the conical form of the chief mass, I suggested that it might have been upheaved amidst these sediments, and have tilted them to the right and left. On recently making a transverse section from the summit on the N.N.W. to the gorge of the Schinbergerach on the S.S.E., I perceived, however, that in parts, the black schists of the flysch passed into a sort of Lydian stone, and that perfectly parallel to the higher zone there were other less elevated peaky ridges of altered millstone grit and sandstone, partially in a state of quartz-rock, with here and there a sort of mica schist. These quartz rocks are sometimes indeed in an amorphous state, and often appear like so many dykes of fused or semi-fused matter running through bands of highly altered flysch limestone. With such appearances therefore on all sides, I could not resist the impression, that the so-called gneiss and mica schist, which I had supposed to be upheaved points of older crystalline rocks, are nothing more than certain courses of the "flysch" which have undergone greater change than the others. Besides, the phænomenon occurs in a highly mineralized zone of the chain; and when I add*, that it is immediately to the north of a grand line of fault, by which the whole system of the flysch and nummulite rocks is brought in its southern flank against the neocomian limestone (see fig. 19) (in precisely the same unconformable relations as at Dörnbirn and Haslach, south of Bregenz), there may be less difficulty in adopting this solution. At all events, the conversion of flysch into gneiss and mica schist is, as before stated, a phænomenon in the Grisons insisted on by M. Studer himself, and a partial exhibition of such metamorphism in the Bolghen may therefore reasonably be admitted.

Prolongation of the Cretaceous and Nummulitic zones of Switzerland and Bavaria into the Austrian Alps.

Taking the strata of Appenzell and those of the Grünten and Sonthofen as types, the practised geologist will have little difficulty in adapting to them the descriptions of the sections of the Alp Spitz near Nesselwang, the banks of the Traun, Kressenberg, Untersberg, Mattsee† and Pancratz, as given by Prof. Sedgwick and myself. Thus, at the Alp Spitz, near Nesselwang, to the east of the Grünten and on the edge of the Bavarian Alps, there is clearly a cretaceous succession, the extent and details of which must be hereafter worked out. But in the meantime, and in reference to our former section, it appears clear that the northern flank of that mountain presents an escarpment in which strata, with fossils of the greensand and gault (if not neocomian), are brought into contact with the same tertiary conglomerates as at the Grünten (molasse and nagelflue)‡. To the south, or towards the Alps, the younger strata of "flysch," &c., are thrown off from these greensands and cretaceous rocks, the most southern of which is evidently a representative of the chalk.

In the section of the Traunstein there is pretty much the same expansion of a system of sandstone and shale and impure limestone with several courses of nummulites, &c. §, as that which has

* M. Boué has described the crystalline mining tract east of the valley of the Iller, which is in truth a prolongation of these masses.

† The fossils which I formerly collected at Mattsee having been examined by M. D'Archiac, are pronounced by him to be *Nummulites Biaritzana* (D'Arch.) (*N. atacica*, Leym.), so common in the Lower Pyrenees, the Corbières, and the Lower French Alps; *N. rotularis*? Desh. (*N. globulus*, Leym.), of the Corbières and the Crimæa; *Orbitolites submedia* (D'Arch.) of Biaritz and the Lower Alps; Operculina, n. s.; Echinolampas, probably the *E. ellipsoidalis* (D'Arch.) of Biaritz; and among the Pectens one species closely resembles the *P. tripartitus* (Desh.) so well known in the tertiary rocks of France. Identifying these beds with those of Kressenberg (see the Bulletin of the Vienna Society, 1848, vol. iv. pp. 267, 269, and Leonhard's Jahrbuch, 1849, p. 109), M. Erhlich has cited from them the *Nautilus lingulatus*, *Clypeaster* (*Conoclypus*) *conoideus*, *C. Bouei*, and the *Micraster pulvinatus* (D'Arch.). As I formerly found such Echinoderms at Mattsee, though at that time they were without names, there can be no sort of doubt of the age of the rock; the Gryphæa of the lowest beds being the only secondary form.

‡ Trans. Geol. Soc. Lond. vol. iii. p. 337. plate 36. fig. 5. The section is so drawn that the tertiary conglomerates appear to be conformable to the cretaceous masses. This is an error.

§ Geol. Trans. vol. iii. p. 338–9, pl. 36. fig. 6.

been described elsewhere, and particularly at Sonthofen. In referring my readers to page 338 of the third volume of the Transactions of the Geological Society, I have only to request them to consider the great group of the flysch, nos. 2, 3 and 4 (Miesenbach to Loheim), as overlying the nummulitic strata nos. 5, 6 and 7, and the whole falls into order with Sonthofen and the Swiss types. To make that instructive section entirely coincide with my present views, I ought to add, that between the northern end of the nummulitic grits and the setting on of the tertiary molasse, the grand fault so often alluded to occurs, and is represented by piles of detritus. The truth is, that the great external fault of the Alps here, as in all the other places cited, inverting the flysch and throwing it off to the south, brings up, against strata of pliocene age, the very oldest or bottom beds of the eocene deposit.

This great fault has, however, been moderate in its operation in Austria and on the south slopes of the valley of the Danube, when compared with the gigantic dislocations that accompany it in Bavaria (Grünten), and in Switzerland (Hoher Sentis and Pilatus, &c.), where even the neocomian limestone, or the equivalent of the very bottom of our lower greensand, is thrust up into lofty escarpments, on the upper surfaces of which the overlying cretaceous and nummulitic groups are pitched in towards the Alps, whilst that neocomian or the oldest formation of the whole succession is at once in contact with younger tertiary nagelflue! Thus whether we appeal to the Austrian, Bavarian or Swiss sections, we perceive (now that we have a true acquaintance with fixed *fossiliferous* base-lines), that there is an ascending order from the point of junction with the younger tertiary, or in other words, that in the valley of the Danube, as in the great valley of Switzerland, or on the shores of the lake of Constance, the underlying members of the series on which in other places the nummulitic group rests, rise up at the very outside of everything alpine, and often throw off the younger portion of the eocene formations into the abnormal position of dipping under the great secondary limestones of the chain.

In regard to the cretaceous group of Gosau, it has been already remarked that it is deficient, both in not possessing any solid limestone with fossils to represent, as in Switzerland and Bavaria, the true equivalent of the white chalk, and also in being void of a distinct nummulitic zone. I have, however, now little doubt, that the sandstone, impure limestone and shale, which there overlie the marls, recognised by their fossils to be cretaceous, are representatives in time of a portion of the nummulitic and flysch series of other parts. In fact, it may be said of Gosau, that the lithological type of the "flysch" there descends not only to the horizon of the inoceramus limestone, to the total exclusion of any limestone to represent the chalk, but also takes possession of nearly all the series of beds which further represent the upper greensand and gault; the first bands of hard and coherent rock being the subcrystalline hippurite limestone, which, like that of the Untersberg, near Salzburg, represents the neocomian formation.

At the Untersberg, the equivalents of the gault, upper greensand and chalk, which repose upon neocomian limestone or hippurite marble, are marls and marlstones, often not unlike malm-rock, and variegated green and red bands, some of them approaching to scaglia, in which Professor Sedgwick and myself found Belemnites and Baculites with *Inoceramus Cripsii* (Sow.) and *Trochus linearis*. Next come sandstone and calcareous grit, with many small nummulites, followed by other strata of sandstone and blue marl, in which other nummulites, with Operculinæ, Dentalia and Serpulæ, are associated with shells having a tertiary aspect. Two or three species indeed of these fossils, such as *Auricula simulata* (Sow.) and *Dentalium grande* (Desh.), have been considered identical with species of the London and Paris basin.

In following the cretaceous rocks from Bavaria into Austria, their upper member or the equivalent of the chalk is no longer to be seen in the form of the white limestone, which is so clear a horizon in Savoy, Switzerland and Western Bavaria. Even at the sections of the Untersberg between Reichenhall and Salzburg, the band containing the chalk fossils is, as above stated, made up of grey, green and red marls and marlstone. In the valley of Gosau, still further to the east, the lithological change is still more decisive; for not only is there no trace of a white limestone, but the group so loaded with fossils, many of which are unquestioned cretaceous species, with many peculiar tertiary-like forms, consists of soft shale and sandy marl, with impure dark-coloured limestones. Reverting however to the sections of the strata above the cretaceous rocks of the Untersberg*, I may affirm, that they exhibit the same general ascending order as at the Grünten, near Sonthofen, and other places, *i. e.* from a true cretaceous zone, (the equivalent of the chalk being in a very different mineral state,) through certain strata of marl and sandstone into masses with nummulites and shells, all of which are unknown in the strata below. On the other hand, it is evident that beds having the characters of the "flysch," are not merely the expanded overlying member of the nummulitic group, but often inosculate with bands of nummulites, and even descend as at Gosau into strata with the true cretaceous fossils. Again, we readily see, that notwithstanding a local dislocation, the highly fossiliferous nummulitic strata of Kressenberg are but a full development of one of these upper bands of limestone, of which I have mentioned many examples.

Not having personally revisited Kressenberg, I can only suggest that the intermixture of a few cretaceous fossils with the acknowledged tertiary types of that locality may be explained† by their having been obtained from the Bergmeister (as at Sonthofen, see p. 206), who may

* The reader must be told, that the true cretaceous rocks with fossils of the age of the gault and chalk, are with great difficulty detected in the slopes between the Untersberg and Reichenhall, owing to the quantity of verdure and detritus which obscure the slopes; but although to a great extent hidden and of no great thickness, they certainly exist in the ravines mentioned by Professor Sedgwick and myself. See Geol. Trans., New Series, vol. iii. p. 346.

† See Trans. Geol. Soc. Lond. vol. iii. p. 344, note.

have collected some of the forms in a truly cretaceous rock. My friend M. de Verneuil, who visited Kressenberg in 1847, has informed me that all the fossils associated with the nummulites are of supracretaceous forms. He has satisfied himself that the matrix of the two sets of fossils is quite distinct, the one containing the gault or greensand fossils being an earthy chloritic sandstone, the other a highly quartzose and ferruginous rock. It is in the latter only, which is surmounted by the flysch, that nummulites occur, including *N. lævigata*, Lamk., and *N. elegans*, Sow., of our London clay, associated with *Orbitoidea*, D'Orb.; *Pygorhynchus Cuvieri*, so abundant in the calcaire grossier of Paris; *Conoclypus conoideus*, which, with other species of that genus, is so frequent in the Alps; and also the *Echinolampas politus*, Ag., common to the Vicentine and south of France.

In a word, there can be no sort of doubt in the mind of any geologist, who has examined the two localities, that the nummulite rocks of Sonthofen are exact equivalents of those of Kressenberg. The flysch at the latter, as at Sonthofen, is thrown in towards the chain, and differs only from that of Sonthofen in the occurrence of a line of fault between it and the beds containing nummulites.

Deferring for the present the general consideration of the fossils of the nummulitic rocks, I may remind the reader that they do not contain any one of the prominently characteristic types of the chalk, such as Ammonites, Belemnites, Hamites, Inocerami, &c. Hence I think that all geologists who classify strata by their animal contents combined with their order of superposition, must admit that the nummulite and flysch rocks of the Alps, Savoy, Switzerland, Bavaria and Austria belong to the older tertiary or eocene age, and can no longer be classed with the cretaceous rocks. The only question, it seems to me, which can be mooted is, where the precise line between secondary and tertiary should be drawn;—for example, whether, as I think, immediately at the base of the lowest band of nummulites, or still lower beneath the flysch-like and greensand beds (*e*) with one or two species of Gryphæa, of which so much has been said. On this point it is sufficient to say, that wherever a true lithological passage and conformable transition occur, the settlement of such line of demarcation must always be somewhat arbitrary.

The opinions of the eminent geologists who have classified the nummulitic and flysch deposits in the secondary rocks, being based upon physical features, I must necessarily defer considering them, until the whole subject of the relations and fractures of this zone be reviewed.

Supracretaceous or Older Tertiary Rocks of the Southern Alps and Vicentine.

The greater part of a century has elapsed since Arduini* expressed his belief, that the deposits of Ronca and Bolca, &c. were of tertiary age, and that Fortis remarked how certain species of fossils from the

* See Arduini's Letters.

Fig. 22.

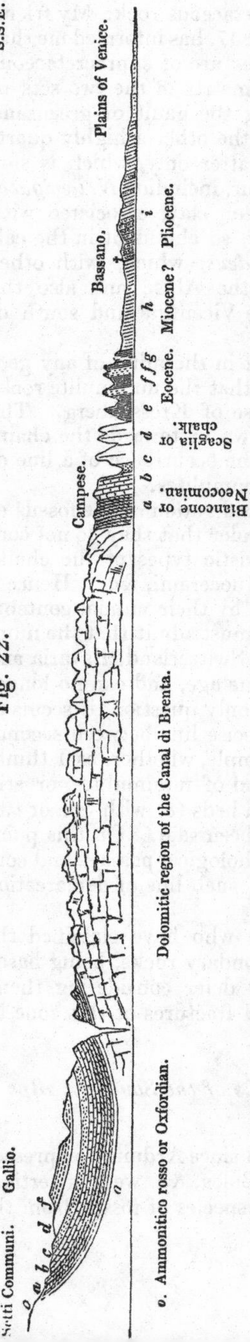

N.N.W. Setti Communi. Gallio. Dolomitic region of the Canal di Brenta. Campese. Bassano. Plains of Venice. S.S.E.

o. Ammonitico rosso or Oxfordian. a b c d Biancone or Neocomian. Scaglia or chalk. e f f g Eocene. h Miocene? i Pliocene.

Val d'Astico resembled forms published by Brander from the London clay of Hampshire. M. Brongniart*, however, first systematically classified these strata as older tertiary, and described their organic remains, whilst our associate Dr. Buckland arrived, about the same time, at a like conclusion in his general survey of the Alps†. But although there could be little doubt, from the absence of cretaceous fossils, and the presence of a multitude of genera having a tertiary facies, that these shelly masses were younger than the chalk, the great desideratum still remained of natural exhibitions of the true relations of these strata to the inferior or secondary rocks. These were the more called for, as some leading geologists considered all the nummulite deposits of the Vicentine to be of cretaceous age. In our rapid survey of parts of the Vicentine in 1828, Sir C. Lyell and myself, having visited the principal localities described by Brongniart as "calcareo-trappéens," found these shelly deposits so commingled with and interrupted by basaltic and other eruptive rocks, that we were unable, any more than our precursor, to detect an order of succession. It was after my colleague had proceeded on his journey to the south of Italy and Sicily, that in returning to England across the Venetian Alps, I obtained the desired proofs of that order, in the clear and instructive natural sections on the banks of the Brenta near Bassano, and again between Possagno and Asolo. I there saw the same nummulitic and shelly beds as those of the adjacent Vicentine, entirely free from igneous intrusions, reposing conformably on the scaglia or the Italian equivalent of the chalk, and passing upwards into strata of younger tertiary age, the whole upraised in lines parallel to the direction of the Alpine chain.

The figures of the remarkable sections near Bassano were published in the 'Philosophical Magazine,' but as that work is in

* Mémoire sur les Terrains de Sédiment supérieurs calcareo-trappéens du Vicentin, par Alexandre Brongniart, Paris, 1823.

† See Annals of Philosophy, June 1821.

the hands of few continental readers, I now reproduce them, with some additions, as woodcuts*, (figs. 22, 23). When these sections near

Fig. 23.

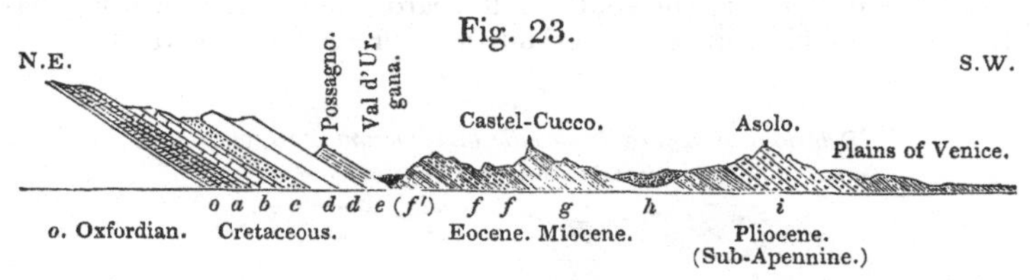

Bassano were described, the new nomenclature of Sir C. Lyell had not been announced, and the groups of shells which there overlie the chalk were simply termed lower and upper tertiary. These two classes of tertiary rocks were shown by me to have been upheaved in parallel lines, and also partially to expose a transition from one to the other. And now that I have revisited the localities, and have examined a much wider range of the Alps, I see more than ever the value of these sections; for as the nummulitic zone is there conformably placed between what I am certain is the true equivalent of the chalk, and a superior zone in which younger tertiary shells occur, the zone so intercalated, and which contains so many true older tertiary forms, must be the representative of the eocene. Nay more, the highly inclined position of the outer or younger tertiary zone would, as I formerly stated, seem to indicate that one of the last great upheavals of the Alps (*redressement*) took place after the accumulation of the sub-Apennine formation. I do not by any means wish to imply that the same elevation which raised the chalk and eocene deposits also raised the younger tertiary deposits. On the contrary, I believe that the latter were thrown up subsequently, but in the same direction as the adjacent older deposits†.

It has already been stated, that a thick mass of compact cream-coloured limestone, with flints and ammonites, called "biancone," now proved by its fossils to be of neocomian age (*a*), reposes on jurassic rocks (*o*), and is surmounted by the whole mass of the scaglia. This scaglia (*d*), containing in parts Inocerami, Terebratulæ and *Ananchytes ovatus*, and being interposed between the neocomian and the group of nummulite rocks with tertiary fossils (*f*, *g*), is demonstrated, like the "sewer-kalk" of Switzerland, to be the equivalent of the chalk. In the headlands between Recoaro on the north

* My last visit to Bassano, Possagno and Asolo was made with the leading members of the Geological Section of the Venetian Meeting of the "Scienziati Italiani," to which I have before alluded. Those who will be at the trouble of consulting my original sections as published in the Phil. Magazine (vol. v. June 1829, p. 401, pl. 5) and those now produced, will perceive that there is nothing essential in the one which is not in the other. The chief alteration is in respect to the flexure or fracture of the cretaceous rocks near Bassano before they come into contact with the nummulitic zone.

† Although upon the small scale the younger tertiary are drawn conformable to the older in figures 22 and 23, there are parts of the intervening tract between Bassano and Possagno where the intermediate sandstones are broken and reversed. Close research may detect an interval between the older and younger tertiary.

and Vicenza on the south, this scaglia is copiously developed, and may be seen in numerous sections underlying the nummulitic limestones. Near Val d'Agno, to the south of Recoaro, the scaglia with its characteristic fossils is directly overlaid (as expressed in this woodcut) by

Fig. 24.

Relations of Lignite to Scaglia and Nummulite Limestone.

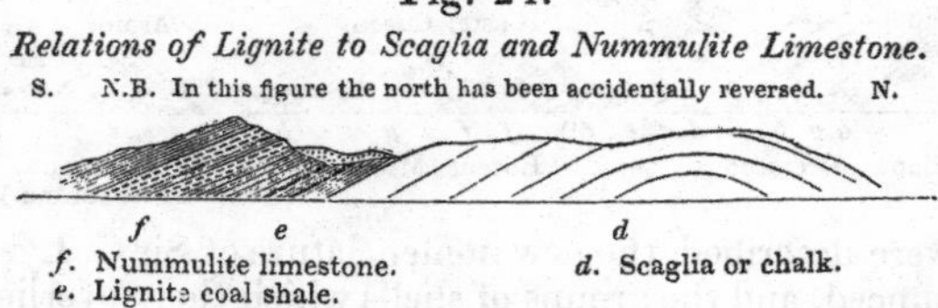

f. Nummulite limestone. *d*. Scaglia or chalk.
e. Lignite coal shale.

seams of coal worked for use in that neighbourhood, which lie in shales that dip away from the older rock, and pass under the adjacent hills of nummulitic limestone. In fact, these coal-beds occupy the same place as those of Entrevernes in Savoy, of the Diableretz, and of the Beattenberg in the canton of Berne (see p. 189). There are, indeed, other localities in this region where the nummulitic rocks are equally characterized by containing lignites or coal, as at Monte Bolca, and at Monte Viale near Vicenza, where it occurs in the escarpment of the well-known coralline limestone of that insulated mount.

In the tract between Vicenza, Schio and Verona, the various sedimentary deposits are so penetrated by different eruptive rocks, whether porphyries, trachytes, greenstones, basalts or serpentines, and peperino, that the dislocations and interruptions are frequent, and the original order of succession with difficulty observed, particularly around Ronca, Montecchio Maggiore, and other localities noted for their organic remains. To the west of Schio, however, and above

Fig. 25.

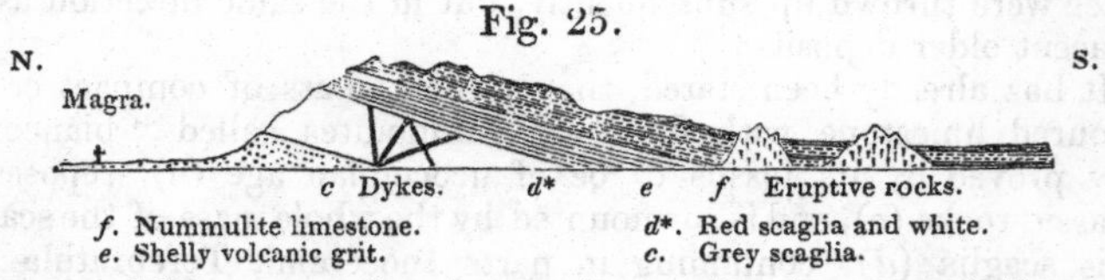

f. Nummulite limestone. *d**. Red scaglia and white.
e. Shelly volcanic grit. *c*. Grey scaglia.

the town of Magra, another instructive section is exposed (fig. 25), the base being composed of the red and white scaglia, in which Inocerami, *Terebratula incurva, Ananchytes tuberculatus* and other fossils occur, whilst the summit is occupied by strong bands of nummulitic limestone. The beds being only slightly inclined, a perfect conformity is observable, as well as a transition from one group to the other. There are here no coal-seams, but towards its upper limits the red fissile scaglia (*d**) alternates several times with basaltic trap tuff, some of the highest beds of which above the scaglia are just as much loaded with nummulites as the hard grey nummulite limestone (*f*) which crowns the hill. The manner in which certain bands of these tuffs are thus interlaminated with the nummulitic strata here, and with other shelly strata of this age in the adjacent tracts, induces me to think that they were volcanic dejections formed contemporaneously with the submarine deposits,—a class of strata now too well known to require further illustration†. At the same time I know that this region also abounds in igneous rocks of a truly eruptive character which have penetrated and cut through the whole of the stratified masses. Examples of both these classes of former volcanic or plutonic action are sometimes to be seen in the same hill, as exhibited in the above woodcut. The chief masses or ledges of nummulitic limestone which thus surmount the scaglia, dip on the whole southwards to pass under the marls, tuffs, sands and limestone of the undulating hills of the Vicentine, and thus the nummulitic limestone is fairly seen to constitute the base of that shelly group, even in a tract much traversed by basaltic matter. But in proceeding to the west and south-west of Schio, the igneous rocks so abound, that a regular sequence, I repeat, is not traceable for any considerable distance. To the north of St. Orso near Schio, indeed, the effects of the intrusion of a great mass of porphyry have been such as completely to invert the strata and to fold back the cretaceous rocks and make them overlie first the nummulitic and then the other and younger tertiary rocks‡, as expressed in the diagram (fig. 26). This point will be

Fig. 26.

Inverted Strata.

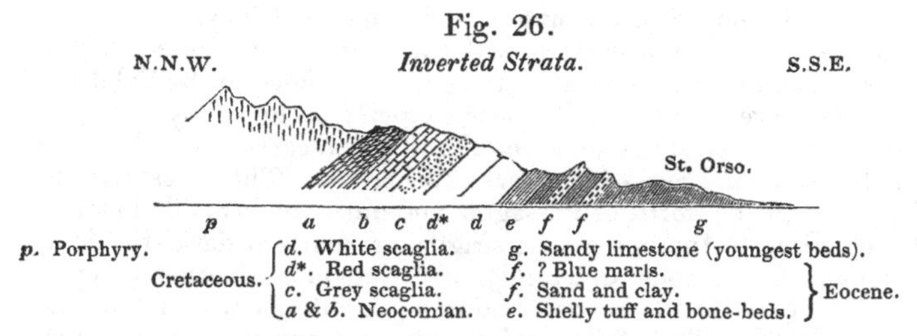

reverted to when the dislocations and inversions in the Alps are considered, and I now proceed very briefly to direct attention to the clear and unambiguous sections of Bassano and Asolo (figs. 22 & 23), which have, in fact, proved to be, what I ventured to suggest so many years ago, the best expositions of the true normal succession from the cretaceous to the tertiary rocks which have anywhere been observed on the flanks of the Alps. (See back, pp. 218, 219.)

On the right bank of the Brenta at Campese, a little above Bassano, the neocomian and scaglia, which range in great undulating terraces

† M. Brongniart has described in some detail these rocks, which he has called "calcareo-trappéens." I only differ from my lamented friend in considering some of his "brecciole" as being contemporaneous with the deposits.

‡ In common with all the members of the Geological Section from Venice, I was exceedingly obliged to my able friend, M. Pasini, for the pains he took to make me better acquainted with the interesting tract around Schio, Recoaro, and the Setti Communi, with which he has been so long conversant. The tract which has given birth to Arduini, Brocchi, Fortis, Mazzari Pencati, Maraschini, and Pasini, may well be considered classical in geology. In this region every variety of dislocation is to be seen with much metamorphism of mineral structure; and yet it is here that the best development of the trias is displayed, as well as a copious series of jurassic, cretaceous and tertiary deposits.

over the summits of the Setti Communi within the chain of the Alps (see fig. 22), are brought down by rapid flexures to occupy, as before said, vertical positions on the edge of the lower country*. To the red and white scaglia (*d*) so placed on both banks of the Brenta succeed sandy marls and stone-bands which form the base of the nummulitic group formerly described. Vertical ledges (*f*) of nummulitic limestone follow. This inclination is continued, as far as can be observed, through the whole space occupied by the city of Bassano; for after passing over the edges of a great thickness of marls, impure limestone, sands, &c. (*g*, *h*), few of which are well exposed, the section terminates towards the flat country on the south in the conical hillock of Monte Grado composed of sandstone, calcareous grit and pebbly conglomerate (*i*), the beds of which strike parallel to the rest of the ascending series, and dip 75° to 80° to the south. I have only to add, that the *Ostrea Virginica* and the shells found in the outermost conglomerate are of pliocene age, whilst the nummulitic and lower masses near the scaglia are of the same date as the older tertiary accumulations of Ronca, Castel Gomberto, &c.

The section from the scaglia of Possagno, on the edge of the Alps, to Asolo, at the exterior of the tertiary series of this region (fig. 23), is much more developed in its middle and upper portions, though the junction of the nummulitic strata with the scaglia, so well seen at Bassano, is not exhibited, the strata having been denuded in the Val d'Urgana. I believe that this valley was formerly occupied by the same slightly coherent strata of shale, marl and green sandstone (*e*), which in the Bavarian and Swiss Alps mark this horizon. The lowest tertiary beds visible to the north of Possagno, and quite conformable in strike and inclination to the underlying scaglia, are marls of darkish colour, occasionally ferruginous and sandy, with fungiæ and other polypes, and many of the fossils so well known at Monte Ronca and in the Berici Hills(*f'*). Then follow calcareous grits and nummulitic limestones (*f*) with *Fusus longævus*, which passing up into hard white courses are surmounted by a yellowish subconcretionary impure sandy limestone with blue fossiliferous marls in which pectens first appear. Next come yellow sandy limestone and calcareous grit with green grains, containing pectens and echini. This mass (*g*) is of considerable thickness, and is very similar to some of the calcareous green sandstones of Switzerland, there associated with the nummulite limestones. Overlying this "glauconie grossière" is a small concretionary mottled dark grey and cream-coloured limestone, loaded with foraminifera, in which nummulites sometimes occur. This rock is well exposed at Castel-Cucco, where it has been largely quarried, and the columns of Canova's church of Possagno are built of it. The strata of marl, shale and sand which succeed to the south of Castel-Cucco (*h*) are badly exposed in low undulating grounds with devious dips; but on reaching the outer tertiary ridge of the Asolan Hills, a good, well-defined order is exposed, first, in an escarpment in which blue marls,

* See Bull. Geol. Soc. Fr. tom. iv. p. 56, 7th Nov. 1842, where M. de Zigno confirms my former view of the general relations of the secondary to the tertiary rocks.

very like the subapennine marls of Brocchi, dip under yellow sandstones and pebbly conglomerates like those of Monte Grado near Bassano (*i*). In the marls are the *Venericardia costata*, *Arca Diluvii*, *Pyrula clathrata*, with species of the genera Murex, Natica, &c., which clearly characterize the blue marls of the subapennine strata; whilst the large *Ostrea Virginica* is found in the overlying yellow limestones and conglomerates. Being aware that M. de Zigno, who has already written on the subject, and has sustained my former views, is about to publish a detailed account of all the species in the tertiary fossiliferous strata of the tract between the Brenta and the Piave, I will not attempt to give palæontological details. I will merely now say, that from the order of the strata and from the fossil shells which our party collected, and also from those we inspected in the museum of M. Parolini of Bassano, I still entertain no doubt that the sections afford an ascending series from the surface of the chalk up into deposits of the subapennine age. M. Ewald of Berlin*, an excellent palæontologist, who in common with M. Leopold von Buch, M. de Verneuil and myself, regarded all the lower portion as eocene, thought that the sandy bands and calcareous grits, which there lie above the nummulitic group, might prove to be the equivalents of the miocene.

But it is with the nummulite group that we are now occupied, and I must leave to local observers the future details and exact delimitation of each tertiary subdivision. It is enough for me to prove that the cretaceous system is here distinctly and conformably overlaid by true lower tertiary deposits, and that the facts which I announced so long ago have now been amply verified; viz. that tertiary rocks, both lower and upper, are in this tract parallel to the secondary rocks, and have been upheaved and set on edge by forces which also affected the adjacent Alps. The lower tertiary group is specially characterized between Bassano and Possagno by containing, in addition to nummulites, *Fusus longævus*, *F. intortus*, *Pleurotoma semicolon*, *Turritella imbricataria*, and a whole suite of shells and many corals completely distinct from those of the chalk, and which are either known tertiary forms of Northern Europe, or species peculiar to the localities. In following the same zone westwards into the Bregonze Hills, and the tracts around Vicenza and Schio, or to the interesting, isolated hill called Monte Viale, it is seen to contain all the species enumerated by Brongniart, among which the following

* Highly valued and esteemed by M. von Buch and all the geologists and palæontologists of his country, M. Ewald seems almost to shun publication. The views which he put forth at the Venice meeting were eagerly caught up by all his auditors. He has since written to me, insisting on the indisputable zoological proofs that these deposits are eocene. He has not seen the species of Gryphæa which I collected in the Northern Alps, and which has been named *G. vesiculosa*, but he contends that the species of this genus known in the Vicentine, and published by Brongniart as *G. columba*, is not a known cretaceous fossil. At the same time he admits that the *Terebratula caput-serpentis* rises from the chalk into the eocene deposits. It is to be hoped that M. Ewald will soon be enabled to resume his journeys southwards, and thus complete a catalogue of all the fossils of the nummulitic group, in which he has already made great progress.

fell under my own notice, *Cerithium giganteum*, *Cerithium Maraschini* (Brong.), (which M. Ewald assures me is the *C. hexagonum* (Brug.) of the Paris basin), *Crassatella sulcata* (Sow.), *Nerita conoidea*, *Bulla Fortisii*. Among the conchifera are the *Pholadomya Puschii?* (Goldf.), the *Cardium Theresæ* of Nice, whilst the *Spondylus cisalpinus* (Brong.) and certain Pectens are as common as in the nummulitic rocks of the Northern Alps. The Echinoderms of this tract are equally decisive of a supracretaceous deposit; for they chiefly belong to the genera Schizaster, Scutella and Echinolampas, which are unknown in the chalk, but which also occur in the nummulitic deposits of Switzerland and Bavaria.

In an excursion through the lofty table-land of the Setti Communi, I saw the lowest tertiary beds, containing *Cerithium giganteum* as well as nummulites, reposing conformably on slightly inclined strata of the red and white chalk or scaglia at the height of about 5000 feet above the sea. This position is expressed in the diagram (fig. 22), which shows how the same movements of elevation and undulation which threw the lower tertiary group into a vertical position on the external flank of the cretaceous rocks at Bassano, had raised fragments of it at Gallio, near Asiago, on the surface of similar rocks amid the summits of the adjacent mountains. Again, in the Kalisberg mountain which overlooks the city of Trent on the east, the uppermost cliffs of sandy yellow limestone, which at a distance weather like dolomites, were found by MM. von Buch, de Verneuil and myself to be nummulitic rocks overlying the jurassic and cretaceous systems*. In these strata we collected the *Nerita conoidea* and *Voluta ambigua*, well-known species of the calcaire grossier of Paris, together with the *Lucina Corbarica* (Leym.), and several species of Echini, including the *Eupatagus ornatus* (Desor), the *Echinolampas subsimilis* (D'Archiac) and the *Pygorhynchus subcylindricus* (Ag.), both of Biaritz, and the *Echinocyamus profundus* (Ag.) of the nummulitic rocks of the Swiss Alps. At Sardagna (Trent) on the opposite bank of the Adige, the nummulitic limestone with Echini, Crustacea, Pectens, and the *Spondylus cisalpinus* so well known at Castel Gomberto, &c., also overlies white inoceramus limestone; thus exhibiting precisely the same succession as at Sonthofen and many other places in the North-western Alps. These nummulitic beds, according to M. Perini, occur also at the height of not less than 7000 feet above the sea, in the peak of Monte Bondone, to the south-west of Trent. The same tertiary deposits, therefore, which form mere hillocks on the south flank of the Alps, and which in some places (Bassano, &c.) are raised conformably into vertical walls, flanking the cretaceous rocks, have been carried up to great altitudes within the chain, where they bear the same relation to the cretaceous formation as the nummulitic rocks and flysch of Switzerland and the Northern Alps.

* In the museum of M. Menapace, of Trent, we observed the *Inoceramus mytiloides* and *Terebratula subglobosa* of the chalk, which that zealous collector, as well as M. Perini, who accompanied us to the Kalisberg, assured us were invariably found under the nummulitic rocks. Most of the fossils above cited were found by M. Menapace.

In resuming the consideration of the deposits which in the Vicentine and adjacent countries overlie the scaglia or chalk, I may add that they sometimes consist of strata, more or less sandy, which alternate with marls and graduate up into sandier bands of greenish calcareous grit. In some tracts so much green earth is disseminated in this series, that near Schio where such is the case, and where the strata have been inverted, as before said, by the porphyry, this band was considered by the older geologists to be the secondary or cretaceous greensand. Professor Catullo has shown, in a recent publication, to what a great extension this zone attains in the Friuli. There it is characterized by a *Pholadomya*, which is scarcely to be distinguished from the *P. margaritacea* of the London clay*. Passing over for the present all the next overlying strata in this section, which the palæontologists of our party believed would prove to be of miocene age when fully examined, I have here only to repeat what I stated in my former memoir of the year 1829, that the highest deposits of the whole series contain many true subapennine shells, and that the beds in which they lie are apparently linked on to those we are now considering.

In regard to Monte Bolca, near Verona, so famous for its fossil fishes, I unhesitatingly affirm that it is of true lower tertiary age. In company with Sir C. Lyell I made sections of it and of the adjacent Monte Postale in 1828, which leave no sort of doubt that the strata are simply continuations of the eocene deposits of the adjacent Vicentine. Marly, whitish and yellowish limestones, occasionally mottled with bluish grey and brown colours, are on the whole subordinate to bands or mounds of peperino, and are also distinctly traversed by dykes of the igneous and basaltic rocks described by Brongniart.

Whilst the latter are certainly posterior, and have in many cases altered the contiguous limestone, the peperino must, I conceive, be viewed as the result of submarine volcanic dejections contemporaneous with the other deposits, the heat attending the evolution of which may have destroyed the fishes of a former well-tenanted bay of the sea, just as on a recent occasion shoals of them were killed on the coast of Sicily when Graham's island arose from the deep. Notwithstanding this abundance of eruptive matter, quite enough, however, of the original sedimentary deposit remains to show, that it is entirely distinct in lithological and zoological characters from any portion of the scaglia or chalk which flanks it on the north. Thus, lignite coal here occurs in the same position as cited in previous pages in the Savoy and Swiss Alps, and at Val d'Agno and Monte Viale in the adjacent Vicentine ; whilst the plants, including dicotyledonous trees, palms, cocoa-nuts, and certain aquatic forms, are pronounced by Dr. Unger to be eocene types†. Nummulites, indeed, occur between the lower and upper fish-quarries, among which I collected the small *N. globulus* and the *N. millecaput*, and with these are associated

* I have mislaid the note which I made concerning the other fossils of this lower tertiary greensand, but besides Ostreæ, I apprehend that it contains a peculiar Gryphæa, like the *G. columba* ? ? of Brongn. of Montecchio Maggiore.

† See also M. Adolphe Brongniart's description of some of the plants collected by his father (Mém. du Mus. d'Hist. Nat. vol. viii. p. 343).

numerous *Alveolinæ*. Whilst mounds of peperino (occasionally however containing nummulites) occupy the upper conical summits of Monte Bolca, overlying the dislocated and variously inclined limestones, true tertiary shells are seen both in the limestones of Monte Bolca itself and of its neighbour Monte Postale. Among these shells are Natica, Fusus, Buccinum, Ostrea and small Avicula, with Teredina closely resembling the *T. personata* of the London clay. We have thus abundant proof of the age of this deposit; but when the fishes are appealed to, they speak the same language still more decisively than those of Glarus in Switzerland. Of the 133 species enumerated and described by Agassiz, many are, it is true, peculiar and unknown elsewhere, but as at Glarus there are genera, and in much greater quantity, which, wholly unknown in any secondary rock, are still living in our seas; viz. Fistularia, Vomer, Torpedo, Lophius, Diodon, Rhombus, Clupæa and Anguilla. The presence alone of many species of *herrings* and *eels* completes the proofs drawn from other sources, that the deposits of Monte Bolca, like all the other nummulitic rocks of the Alps, must be completely severed from the chalk, and be considered a true lower tertiary formation.

Most geologists must, indeed, have been disposed to adopt this conclusion from the tabular arrangement of Agassiz, who, while the subject was still a matter of doubt, prudently placed the ichthyolites of Monte Bolca, together with those of Monte Libanon, as a special group between the cretaceous and tertiary deposits. I now, however, revert to the old opinion of Fortis, and definitively, I hope, class the Bolca deposit as a true lower tertiary rock.

I may terminate this portion of the memoir by saying, that when we compare the Vicentine and Veronese eocene deposits with the nummulitic rocks of the Savoy, Swiss and Bavarian Alps, we find as much assimilation as can be expected to occur in deposits of the same age, but of dissimilar composition, which lie at some distance from each other, and have manifestly been separated by intervening lands. In both, the true equivalents of the chalk are overlaid by limestones, in which some of the same species of nummulites appear interstratified with and overlaid by deposits in which are many of the same shells; whilst the most striking parallelism is marked by the abundant echinoderms of the two regions—all quite distinct from those of the preceding æra. In short, the deposits on the south as on the north slopes of the Alps are proved, by their organic remains and superposition to rocks containing chalk fossils, to be of the lower tertiary age, provided the groundwork of the classification previously adopted by geologists be not entirely changed.

In many natural sections, where the disruptions so frequent in this chain have not interfered, the evidences are complete as to a former continuous deposit from the surface of those strata in which any cretaceous fossils are discernible, through a vast series of strata in which all the vestiges of life belong to a new æra. What then can these nummulite deposits, whether in the Vicentine or in the Swiss Alps, be, but true eocene? If there be geologists who are not swayed by the evidences of organic remains only, still they must surely be influenced by the existence of a great, *conformable* and *continuous* succession of finely laminated strata, the deposit of which being clearly proved to begin after the accumulation of the limestones with true chalk fossils, has gone on uninterruptedly during long ages. The united group of the nummulite limestone and flysch of the Swiss Alps, as well as the great nummulitic and shelly accumulations of the Vicentine, are indeed more stupenduous monuments to mark the lapse of *time* than any of the so-called eocene deposits in Northern Europe. This phænomenon of a fuller eocene development, at least of all its lower part, in Southern Europe, is quite consonant with the facts elicited by the geologist. In Northern Europe a hiatus is very generally seen between the surface of the chalk and the lowest eocene, occasioned doubtless by very considerable disturbance at that æra. In numberless places the surface of the chalk has been abraded by the action of tumultuous waves, and the strata have been dislocated before the tertiary strata were accumulated thereon: not so originally in the Alps. There, the submarine deposits having in many parts been continuous throughout both periods, we are necessarily presented (where subsequent dislocations have not obscured them) with a grander series of strata. In regard to the enormous thickness of "flysch" which overlies the zone of nummulites and other recognizable fossils, and in which very little of organic form, save fucoids and a few fishes' teeth and scales, and an occasional cast of a shell, have been detected, we can scarcely say more than that, from the intimate association and intercalation of these rocks with nummulites, we must presume that they were simply the copious accumulations of a deep sea of that æra in which animal life was scarce. It is however to be noted, that the well-preserved ichthyolites of the Glarus slates, which unquestionably occur in one of the lower bands of flysch, are highly important evidences, and not less so that they are accompanied by the bones of a bird and a tortoise. The fishes of Monte Bolca, their position and their association with nummulites, enjoin still more forcibly the same conclusion. The fucoids of this deposit are indeed of little value in geological classification. For although in the Swiss and Bavarian Alps they mark, as far as I know, the upper portion of the group we are now considering, there are forms said to be similar in the Italian Alps which occur in the grey or lower chalk beneath the red scaglia. And this is just what we might expect; it being almost an established law in the distribution of organic remains, that the higher the organization the more neatly defined is its stratigraphical horizon. Vegetables of so low a class as fucoids, and so adapted for enduring physical changes, may therefore have continued to live on in spite of those grand mutations which may have often interfered with animal life.

It may be objected that the "flysch" of the North-western and Austrian Alps is not obviously displayed in the same mineral form on the flanks of the Southern and Venetian Alps. But even there the yellowish and green sandstone, and bands of marl and schistose limestone which are associated with the nummulite zone, may well be viewed as representatives of the North Alpine flysch. It is, in fact, from the identity of the rocks, and the belief in a similar position of the Italian macigno (upper) and the flysch of the Swiss, that M. Studer has recently styled the latter "macigno alpin." In treating of the Carpathians and Apennines it will, however, be shown to what a limited extent the "grès des Carpathes," and the Italian "macigno," are to be identified with the nummulitic or lower tertiary flysch; for in both these regions it happens, that the same lithological type of sandstones (often green) pervades vast thicknesses of strata, some of which are of upper secondary and others of lower tertiary age.

On the younger Tertiary Rocks of the Alps, and on the extent to which they represent the Miocene and Pliocene of Geologists.

In all parts of the Northern Alps there are evident signs of a marked interval between the last-formed strata of eocene age and the next overlying deposits, which every one has admitted to be tertiary. In contrast with the apparent conformable superposition of the overlying tertiary strata to the eocene on the flanks of the Venetian Alps near Bassano, already alluded to, and in parts of Italy to which I shall afterwards advert, the general phænomenon along the northern edge of the Alps, is that of a grand dislocation between such masses. In other words, it is clear that between the upper portion of the "flysch," and the lower portion of any tertiary formation of subsequent date, there is so great a break and unconformity as quite sufficiently to account for the absence either of the uppermost eocene, or of the lower part of the miocene of other countries.

Professor Studer, who has so long and so minutely studied the molasse and nagelflue of his native country, has as yet in vain sought for any section which exhibits a physical connection between the base of these deposits of molasse and the upper portion of the strata we have been considering. Thus dissevered from pre-existing strata, the molasse and nagelflue conglomerates are constantly thrown up at all angles of inclination, not only to verticality, but beyond it; and on lines usually parallel to the direction which has been impressed on the pre-existing masses of the chain, viz. from W.S.W. to E.N.E. The manner in which many of these tertiary conglomerates and molasse have been so placed against the flanks of the chain will be presently considered. In the meantime, whilst I acknowledge my incapacity to work out the subject completely, let us see what can be gathered from fossil evidences respecting the true age of these deposits.

In Styria * there is, indeed, a general ascending series, from a base

* The account of the tertiary deposits of the Styrian Alps, by Sedgwick and self (Trans. Geol. Soc. Lond. vol. iii. 2nd series, p. 382), has recently received a great addition in the description of their fossil plants by Dr. Unger of Grätz. (See Leonhard and Bronn, Jahrbuch, 1841, p. 505, and Journ. Geol. Soc. Lond. vol. v. Part 2. p. 11.) That author enumerates nearly 150 species from one bed only of lignite at Parschlug, all of which are of lost forms. Besides many Dicotyledonous trees of genera common in Europe, there are genera which require a climate as warm as South America, whilst others resemble the fauna of the United States and table-land of Mexico. On the whole, Dr. Unger believes that these plants,

with partial conglomerates, the whole reposing upon older rocks, and dipping away under the younger deposits of the adjacent lower countries. But when we pass to the north flank of the Alps, particularly in Bavaria and Switzerland, the physical relations are manifestly different.

In speaking of Switzerland I must not only refer to the well-known and excellent work of M. Studer on the Molasse, but also to the valuable additions to it recently made by M. Arnold Escher de Linth *. In the former we have set before us numerous derangements of this great deposit, and also the variations in its composition in different tracts. In the grand pebbly accumulations of the Rigi, for example, several thousand feet of which are clearly exposed, there are pebbles of granite and porphyry whose parent rocks † are now wholly unknown to the mineralogist in the Alps. At the same time it is clear, that the chief heaps of such materials have been derived from the well-known adjacent ridge of secondary limestone, mixed up with an extraordinary quantity of "flysch," which rock has also afforded materials for a large portion of the calcareous sandy matrix of the nagelflue. M. Escher points out that this great system of nagelflue and molasse is divisible into three zones. The lowest visible portion of the inferior zone is exposed along a great axial line, which, according to M. Escher, passes from near Rheineck on the north-east, by Herisau, Watteville, Jonen east of Rapperschwyl, on the north bank of the lake of Zurich, and Hutten on the south-west. Thence it runs between the lakes of Egeri and Zug immediately to the north of the city of Lucerne, whence it is presumed it may be followed further to the south-west, to the west end of the lake of Thun and the valley of the Sulg. Along this line molasse sandstone is seen in vertical or highly-inclined positions, throwing off overlying conglomerates of enormous thickness. If the masses of nagelflue which constitute the Rigi mountain near Lucerne, and the still loftier Speer (figs. 12 & 14, pp. 195, 200) near Wesen be included in one group, their thickness must be enormous, certainly exceeding 6000 or 8000 feet. This axial line trends from W.S.W. to E.N.E., and is, I would remark, perfectly

which I examined in his company in the museum of Grätz, bespeak a Mediterranean climate and a miocene age. It is eighteen years since I furnished M. Adolphe Brongniart with the plants of the Häring tertiary coal deposit in the Tyrol.

* For M. A. Escher's account of the molasse of Eastern Switzerland see Mit theilungen der Naturforschenden Gesellschaft in Zurich, No. 7, May 1847. In this memoir M. Escher states, that although a powerful deposit of marine molasse (not less than 1000 feet thick near Berne) is interpolated between the lower and upper freshwater molasse and nagelflue, he is unaware of any zoological distinction in the two last-mentioned members of this great series. *A warm climate, which permitted the growth of palms and large Cycadeæ, seems to have prevailed during the whole of the molasse period*, and the species of Helix, Lymnea, Planorbis, Melania, appear to be the same in the strata above as well as in those below the marine molasse.

† Professor Studer believes that the parent granite, from whence such pebbles were derived, protruded along the great line of dislocation between the molasse and the chain, and was lost by subsidence *en masse* when the great accumulations of nagelflue were formed.—(Letter to myself.)

parallel to the great band of cretaceous rocks, nummulite and flysch before described, to which the strata of the nagelflue and molasse are entirely unconformable. Nor does this line of dislocation cease at the eastern end of Switzerland. It continues, as before stated, in the same direction, from near Bregenz to Immenstadt in Bavaria, where it affects the huge tertiary masses, often vertical and sometimes dipping both to the north and south, in which Professor Sedgwick and myself have described several transverse sections between Bregenz on the south-west, and the subalpine ridges south of Munich on the E.N.E., in which micaceous sandstones with marls, shales, conglomerates and courses of lignite occur, as in Switzerland. In some Bavarian strata of this age we found freshwater shells, *i. e.* Cyclades and Potamides, mingled with marine forms *. On the whole, however, we detected so very few fossils in these vast accumulations, that, simply connecting these rocks with the molasse and nagelflue of Switzerland, we then said that whatever conclusions Professor Studer or other geologists might establish, by help of fossils, respecting the Swiss formations, might be extended to a portion of the newer Bavarian deposits. Now, in what is called the lower group, particularly as seen in the canton of St. Gallen and along the axial line above cited, no trace having been found of anything organic except lignite with terrestrial plants, and land or fluviatile shells with bones of extinct land quadrupeds, M. Escher justly considers it to be a freshwater formation.

Whatever may be the dimensions of the lower (freshwater and estuary) member of this series, it is overlaid by molasse, sandstone and marls of considerable thickness, which contain a great variety of marine species †. I submit a collection to the Society which I obtained from Professor Deicke at St. Gallen, near which place they abound. In examining the strata there, in company with that gentleman and Professor Brunner, I perceived that the shells chiefly occurred in beds of sandy, micaceous blue marls, which alternate with sandstones, and are intercalated with large accumulations of pebbly conglomerates. The following may be enumerated as among the characteristic fossils which occur at St. Gallen, but more complete lists must hereafter be given; viz. *Solen vagina,* Linn.; *Panopæa Faujasi,* Menard; *Cardium multicostatum,* Broc.; *Venerupis eremita* (*Venus,* Broc.); *Venericardia Jouanetti* ?, Desh.; *Pinna nobilis,* Broc.; *Pecten scabrellus,* Broc.; *P. latissimus,* Broc.; *Conus turricula* (Broc.), with other species of that genus; *Turritella terebra,* Broc.; *T. vermicularis*; *Pyrula reticulata,* Lamk.; *Natica canrena* (*Nerita,* Broc.); *Phorus agglutinans,* Lamk., as well as species of the genera Pholas, Venus, Cardium, Dentalium, Serpula, Balanus, &c.‡

The sections of St. Gallen (as pointed out to me by Professor

* These beds are described in Geol. Trans. 2nd Ser. vol. iii. pp. 326, 329, 370.

† For the general relations of these freshwater and marine strata of the molasse see the woodcut, fig. 14, p. 200.

‡ Whether these St. Gallen fossils be called older pliocene or younger miocene is immaterial to me, as I only seek to show that among them are numerous existing marine species. (See subsequent observations.)

Deicke) exhibit the strata with marine shells intercalated between freshwater deposits, which contain the *Melania Escheri* (Merian) and *Planorbis hispidus*, Pupa, Melanopsis, and small Potamides, with seams of lignite, &c.

The enumeration of the fossils of the marine molasse of St. Gallen, though far from being complete (not more than a third of the species I saw are mentioned), is I think sufficient to prove that these beds are of nearly the same age as the blue subapennine marls of Italy, and therefore of what has been called the older pliocene age. The marine shelly beds of the molasse in the canton of Berne, also low in the series, are equally referred by Professor Studer to this age; for although the shells there are neither so well preserved nor so numerous as at St. Gallen, the presence of the *Panopæa Faujasi*, *Pecten laticostatus* (Brod.), *Cyprina Islandica*, *Tellina tumida* (Brong.), all characteristic shells of the subapennine deposits amidst those which are recognizable, leaves little doubt on the subject. In Berne, as in St. Gallen and Zurich, the marine beds in question surmount (according to Professor Studer) a widely-spread lower freshwater deposit.

In the canton Vaud, where remains of tortoises, crocodiles and extinct quadrupeds occur, the order of superposition and relations of the different masses of the molasse are obscurely seen, particularly in the undulating region between the lakes of Neufchatel and Geneva. Still it is right to observe, that in the environs of Vevey, where molasse and conglomerate abound, no traces of any marine remains have been found; the only-fossil indeed known there being a Palmacites of some size, detected by M. Collon *. There the tertiary conglomerate and molasse are truncated, and with an inverted dip (fig. 4, p. 182) seem to dip under the adjacent secondary rocks as in the diagrams (figs. 12 & 14, pp. 195, 200), though here they are in contact with rocks of the age of the Oxfordian Jura.

That marine strata overlie freshwater conglomerates, is indeed clearly perceived in the environs of Chambery and other parts of Savoy. The Canon Chamousset accompanied me to sections, where a conglomerate made up of the detritus of the adjacent neocomian limestones contains freshwater shells and lignite. In that tract, where all the intervening strata, representing the gault, upper greensand, chalk, nummulitic limestone and flysch, are absent, the freshwater conglomerate reposes at once on the secondary neocomian limestones from whence its materials have been derived, and passes upward into the marine molasse, as exposed in the woodcut (fig. 5, p. 184).

This lower freshwater accumulation in Savoy is not less than 1000 feet thick. Its lowest beds consist of limestone conglomerates followed by red marls and marlstone with green veins and spots, and occasional gypsum. Then follow other calcareous pebble bands, containing subordinate courses of marly limestone with freshwater shells.

* M. Blanchet of Lausanne has a rich collection of fossils from these fluvio-lacustrine deposits of the canton de Vaud. He believes that these mixed deposits are of different ages, each varying according to its proximity or remoteness from the chain of mountains from which it was washed into the bay by rivers (see his Supplement).

The latter are surmounted by marly sandy beds, approaching in character to molasse, which gradually pass up into the true marine molasse.

The marine molasse of the cantons St. Gallen and Zurich dips to the N.W., and is clearly surmounted by enormous accumulations which constitute the upper nagelflue, and throughout which nothing but terrestrial or freshwater remains have been detected, the species, of the genera Melania, Helix, Planorbis, Lymnea, being apparently undistinguishable from those of the nagelflue and molasse beneath the marine strata. It is probably this great upper member which is for the most part thrown into the remarkable inverted position exhibited in the diagrams figs. 12 & 14. In a portion of this upper member at Kapfnach, and in the Albis Hills near Zurich, are found freshwater beds, in which Helices and seeds of Chara occur together with the bones of *Mastodon angustidens, Palæomæryx, Orygotherium Escheri, Chalicomys Jägeri, Cervus lunatus, Hyotherium medium, Rhinoceros Schinzii,* all species recently described by M. Herman von Meyer. In the same deposit, leaves of Acer as well as parts of palmaceous plants are seen*.

Again, molasse and conglomerate occur in still higher positions; *i. e.* in the summits of the ranges near Zurich, where the pebbly beds are very cavernous, and have given rise to the name of "löchrige Nagelfluh;" but no characteristic organic remains have been found in it.

In following the surfaces of these vast accumulations as they recede from their dislocated and highly inclined positions on the flanks of the Alps into the great trough which extends up to the Jura, we find the beds becoming more and more horizontal, in which position they range up to the edges of the latter mountains. The same order of strata is however observable, and every here and there we see—notably near Baden in Switzerland—courses of marine shelly marls and sands charged with the same group of subapennine fossils† and covered by freshwater nagelflue.

The vegetable remains of the molasse seem all to be referable to a warm or Mediterranean climate, and they are all extinct species. To this consideration I shall presently revert.

* The *Mastodon angustidens* occurs at several other localities, viz. Buchberg, Elgg, Greit, &c. The *Rhinoceros incisivus* is found at Elgg, and the *Rhinoceros Schinzii* (Herm. v. Meyer) was extracted from nagelflue at Bolingen, near the foot of the Albis, where it is associated with *Unio Escheri* and extinct species of Paludina, Melania, &c. Molasse fossils, including tortoises, are also in force at Winterthur. This upper group of molasse with mammalia is clearly separated from the horizontal older alluvia of these regions, of which there is a fine example at Utznach, in which the *Elephas primigenius* or mammoth occurs, with land and freshwater shells, and pines, and other vegetables of existing forms.

† See the list of these fossils in the excellent monograph of the Baden country, by Professor Mousson of Zurich, "Geologische Skizze der Umgebungen von Bade im Canton Aargau von Alb. Mousson, Zurich, 1840." In this work the reader will find a very instructive tabular arrangement of all the jurassic and underlying rocks, which are very closely paralleled by fossil species with the oolitic deposits of England.

Freshwater Deposits of Œningen.

In following the surface of the uppermost beds of the nagelflue and molasse from the lofty hills which flank the chain of the Hoher Sentis, &c., the formation as it spreads over the lower grounds, extending from thence, and from the lake of Zurich to the lake of Constance and the Rhine, is chiefly characterized (where fossils and lignites have been detected) as a great terrestrial or estuary deposit. On the right bank of the Rhine, between Constance and Schaffhausen, the celebrated freshwater deposit, which I visited for the third time, has, it still appears to me, been formed in a depression of pre-existing molasse and nagelflue*. In revisiting this locality I was anxious to see what discoveries had been made, and what influence they might have, in conjunction with the recent description of the fossils, on the conclusions respecting the age of that formation which I formerly entertained. In regard to its overlying position I am happy to say that my former general view is supported by M. Studer, M. Escher, and all the Swiss geologists; viz. that these freshwater sands, marls and limestones are younger than the chief masses of molasse and nagelflue of Switzerland. As in my previous communication a very small woodcut only was given, I beg to annex another which better represents my present ideas.

Fig. 27.

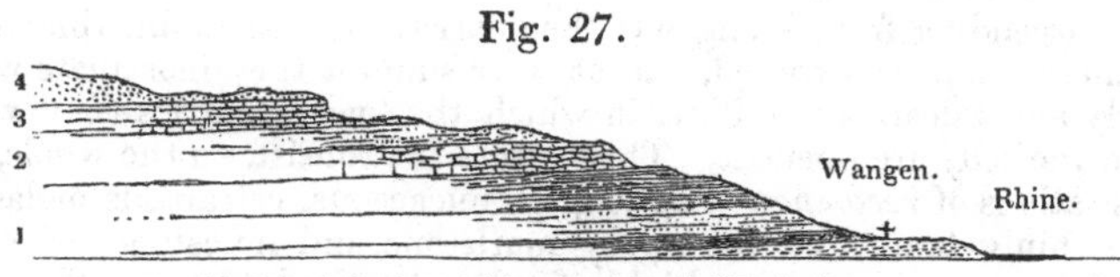

4. Marls and detritus (with volcanic tuff of Escher).
3. Upper quarries of freshwater limestone.
2. Lower quarries of freshwater limestone.
1. Regenerated soft molasse (marine molasse of Escher).

The area over which the sandy marls, marlstone and limestone of this deposit extend, is of much greater dimensions than the spots where quarries have been opened and wherein the fossils have been found. This area, as far as it can be traced, is of an elongated elliptical form, extending with the Rhine from Berlingen, on the right bank of the river, to Wangen and Œningen near Stein on the left bank, a distance of not less than ten miles from east to west. This is inferred because freshwater shells have been found in the soft recomposed sandstone of Berlingen, which rock is of the very same character as that which

* On this occasion I was accompanied by Professor Brunner. For my previous description of Œningen see Trans. Geol. Soc. Lond. 2nd Ser. vol. iii. p. 277. In the little woodcut there given the surrounding molasse and nagelflue were indicated by inclined lines, though I then knew perfectly that in this tract such strata were not there inclined. These lines were only inserted to mark more strongly my belief that such rocks, *so highly inclined in the neighbouring country*, were of age anterior to the overlying marls and limestones of Œningen. See also the account of this deposit by M. Escher von der Linth, given by Herman von Meyer in his Palæologica, 1845, and my observations thereon, Journal of the Geol. Soc. Lond. vol. iii. p. 54.

forms the bottom of the so-called Œningen deposit (1). The ancient lacustrine expanse may indeed have occupied much of the broad valley now filled by the Rhine and the Unter See or lower lake; so that it is difficult to define its former limits on the E. and N.E.* To the south, however, and to the west and north-west it was manifestly bounded by hills of hard pre-existing nagelflue, whose summits are surmounted by erratic blocks only. No one can ascend the indestructible rock of nagelflue from which the castle of Hohenklingen overlooks the town of Stein, and then examine the edges of the contiguous freshwater accumulation, without coming to this conclusion. It is, indeed, evident that the lacustrine deposit was bounded by these hard rocks. The lowest beds of the Œningen basin, as seen in the ravines between Stein and Wangen, and in the lower terraces under the plateau of fossil limestones and marls exhibited in the preceding woodcut, are incoherent, micaceous, light-grey sands, with an occasional concretion (1) fig. 27. They are, in fact, regenerated molasse, and have been compounded out of the hard dark-coloured molasse building-stone, to which they have much the same resemblance, as the sands on the shore of a lake to the sandstone cliff on its sides from whence they have been derived. This is, I repeat, exactly the same soft stone as that which recurs at Berlingen, between Constance and Stockhorn, on the opposite bank of the Rhine, and where freshwater shells are found in it.

In ascending from Wangen to the quarries, a considerable thickness of these sands is exposed, and at their summit they inosculate with marly and calcareous courses, in which the lower quarries (now very little worked) are opened. Their strata (2) consist, on the whole, of alternations of *recomposed,* light-grey, micaceous, calcareous molasse, with thinly laminated, dark-grey marlstone and limestones of conchoidal fracture, which are highly fetid under the hammer. Though of irregular persistence and somewhat broken, these beds (the upper part of which is ferruginous) incline slightly to the west, or away from the valley of the Rhine to which they present their edges, and by which inclination they are carried under all the limestone and marl of the plateau. Among the fossils which they have afforded are the *Palæomœrix* of V. Meyer, together with portions of tortoises; but owing to the concretionary form of the beds and the irregularity of their composition (*i. e.* sand and marlstone inosculating), the fossils are neither so well preserved, nor so much sought after, as in the overlying quarries of flat bedded character.

Rising gently along the inclined surface of the plateau above the lower quarry, the substrata around the dome-shaped ground of Solenhofen are seen to consist of similar rocks passing upwards into marlstones or limestones, which at the distance of about three-quarters

* M. A. Escher de Linth makes the freshwater beds extend northwards by Schienen to the valley of the Aach. I did not revisit that portion of the ground, but I have perfect confidence in his section. The recent discovery, however, of freshwater shells in the underlying band at Berlingen (since M. Escher wrote) decides the nature of the band (1) of my section, which he termed with doubt "Meere's? molasse." (See Fauna der Vorwelt von H. v. Meyer, 1845, p. 49.)

of a mile from the lower quarry are fully displayed in the upper quarry*, the descending order in which is as follows:—

	Ft.	In.
Soft and decomposing bluish grey and white marls used as brick earth, the lower portion consisting of courses from two to eight inches thick, of finely laminated marlstones with very thin laminæ of chert, about .	20	0
Soft bed in which the tooth of a mastodon has been found	1	6
Fish-bed (marly limestone), fishes abundant	0	2
Insect-bed (very finely laminated)	0	2
"Kleine und grosse Moden," stone bands with a few fishes	4	0
"Salamander Platten," in which the *Andrias Scheuchzeri* was found; fishes rare	0	5
"Schildkrot schicht," or tortoise-bed, in which the *Chelydra Murchisoni* (Bell) occurred	0	6
Shale or marl, varying from two or three inches in one part to two feet in others	1	0
"Diehl Stein," or plank-bed, so called because it breaks into long thin board-like flags	1	0
Fox-bed, *i. e.* the marly limestone enclosing the *Galecynus Œningensis* ...	0	4
Fish-bed with numerous fishes, frogs, and several small quadrupeds.........	0	6
"Kessel Stein," or bottom beds of the quarry loaded with plants and the freshwater shells "Anodonta"	1	2

High as it may be in the geological series, and posterior as it certainly is to the marine strata of St. Gallen and Baden with certain existing species of sea shells, the Œningen deposit is not, however, as I formerly supposed, a link between extinct and existing nature. Indeed, whilst I expressed that opinion, I contended that stupendous changes had occurred since this lacustrine matter was accumulated. I showed to what a depth the valley of the Rhine had been subsequently excavated, and how the drift, erratic blocks, and löss had afterwards been deposited; but judging from the best opinions I could then obtain from naturalists respecting the characters of the animals, whether quadrupeds, fishes, shells or insects, or from the plants, I was led to think that they very nearly approached, and in some cases were undistinguishable from, living forms.

More precise researches, however, lead to a very different conclusion. Amidst the multitude of well-preserved fossils, not one, it is now said, is strictly identifiable with an existing species. The closest analogy, indeed, exists between the manner in which the animals and vegetables have been entombed in the mud of this former lake and that which would still prevail. The fossil insects Blatta and Nepa are there found, as I formerly said†, collocated with remains of the

* On this occasion we were so fortunate as to find the present proprietor of the quarries, M. Barth, busily directing his workmen, and as he has made researches for many years, I took down the description of each stratum from him. M. Barth having been unfortunate in trade now devotes himself exclusively to the extraction of the rarer fossils, and in preparing suites of them for sale. M. de Seyfried of Constance possesses the most perfect of the collections of the Œningen fossils with which I am acquainted, all found since I last visited that country. In it I observed five noble specimens of *Andrias Scheuchzeri* (Homo diluvii testis), *Lagomys Œningensis*, *Chelydra Murchisoni*, and another species of tortoise undescribed; and among many splendid fishes an eel three feet long, the *Coluber Oweni*, the tooth of the *Mastodon angustidens*, &c.

† Trans. Geol. Soc. London, 2nd Ser. vol. iii. p. 286.

leaves of the same genus of tree on which they still live; but the species are distinct from those now prevailing. On this point I quote the opinion of Professor Heer of Zurich. That zealous entomologist assured me, that out of 120 species of Coleoptera, 40 species of Neuroptera and 80 species of Hymenoptera (60 of the latter belonging to Formica), he has not, after the most rigid microscopic comparisons, been able to detect a single form, either aquatic or terrestrial, which can be identified with species now living in any part of the globe. Some of them, indeed, make close approaches to species now living in America and the Mediterranean, including Algeria, and some genera (at least six) are entirely new*.

Professor Agassiz classes the fishes of Œningen much in the same category, and the same may be said of the numerous quadrupeds, whether those so elaborately and well described by M. Herman von Meyer, or the extinct form of the Viverridæ named by Professor Owen *Galecynus Œningensis*, or the "Fossil Viverrine Fox of Œningen†." Even in regard to the plants, it does not appear that any can be identified with living forms; for although M. Göppert has said that he can discover no difference in one case between the cone of a pine of Œningen and the cone of the living *Pinus sylvestris*, he admits that without further evidence as to the glands and leaves, no proof can be obtained that it is not an extinct species.

Such being the facts, how are we henceforward to classify with certainty, tertiary deposits which have been formed *on land*, in relation to those which have been accumulated *in the sea*? In the latter, or the marine Swiss molasse, we find that strata formed anterior to the Œningen deposit contain shells of the subapennine æra, many, or some at all events, of which are now living in our seas‡; whilst the land and fluviatile animals of posterior date are *all* distinct from those now in existence. In reviewing the molasse and nagelflue as a whole, the evidence, as far as it goes, teaches us, that the formation was in many tracts almost entirely formed by rivers or in lakes; whilst in other parts, as near Berne and St. Gallen, there were powerful intercalations of deposits formed in bays of the sea. If then we consider the whole as a connected series, and admit that in the lowest as well as in the highest strata, and even up to the regenerated molasse and marls of Œningen, the land remains belong to extinct species, still we

* Professor Heer's monograph of the fossil insects of Œningen will, I doubt not, interest all entomologists as well as geologists, by the knowledge it exhibits of every analogy and comparison which can be set up between these fossils and the living forms of insects. Professor Heer intends to describe in a subsequent work the insects of Aix en Provence and other tracts.

† The animal collected by myself, and described as a fox by Mantell, is now named by Owen *Galecynus Œningensis*, or the "Fossil Viverrine Fox of Œningen." See Journal of the Geol. Soc. London, vol. iii. p. 55, with anatomical woodcuts.

‡ I here conform to the more generally received opinion concerning pliocene marine shells as advocated by Sir C. Lyell and M. Deshayes. M. Cantraine indeed believes that nearly all the true pliocene or subapennine species are still living (see Malacologie Mediterranienne et Littorale, Acad. de Bruxelles, tom. xii. des Mém. 1840). On the other hand, however, it is right to state, that M. Agassiz contends that no animal having the exact form of a fossil tertiary mollusk is now living in our seas.

have the remarkable fact, that in the subordinate marine masses many of the shells are living species.

This discrepancy in the evidences drawn from terrestrial and marine sources has already created divergent opinions respecting the age of strata among naturalists. Thus, judging from the vertebrata found in the older freshwater deposits of the Rhine and other parts of Germany, where marine evidences are wanting, M. Herman von Meyer would class as eocene that which other geologists call miocene, and he has naturally referred to the miocene age those very Œningen freshwater and terrestrial strata so charged with lost types, but which, as I now assert, were formed after the accumulations in which pliocene and living marine fossils occur.

This persistence of marine forms during a period in which a whole terrestrial fauna became extinct—a period it will be recollected when the proportion of the known remains of the land in reference to those of the sea was infinitely larger than in earlier times—may lead us to be cautious in deciding on the age of a secondary rock by the mere characters of its fossil vegetables (see p. 178). At all events, the contents of the upper tertiary deposits of Switzerland compel us to admit, that in any classification of a terrestrial formation by the more or less prevalence of existing types, not even the youngest of those Swiss strata at Œningen can be termed miocene or pliocene. So completely, indeed, do all its imbedded terrestrial animals seem to belong to lost types, that we have not yet even authority to call them eocene, although in reference to marine deposits they have been formed in part out of the detritus of the marine eocene Alpine rocks! In rendering our science exact, we must, therefore, I apprehend, classify strata deposited in fresh water or on land separately from those of submarine origin. In reference to the tertiary æra, we can only speak of the former, as *older* or *younger land* formations; since it is manifest, that (without a total disregard of the meaning of the words) we cannot apply to them the terminology employed to designate the tertiary *marine* stages*.

Dislocations in the Alps.

The previous pages having been chiefly devoted to the detection of the order in which the formations have been accumulated, I now invite attention to some examples of those grand phænomena of contortion and fracture of the strata which specially characterize these mountains. By whatever causes produced, these derangements are so great, that geologists accustomed to work in less troubled regions could scarcely have ventured to hope, that the Alps would have been found to explain any portion of the *succession* in the earth's deposits, still less that they should contain, as I have endeavoured to show, certain *links* to connect the secondary and tertiary rocks, which, if

* The commingling of lost types of large terrestrial animals with those of species scarcely distinguishable from our own in the rich tertiary deposits of the sub-Himalaya chain, is, also, a splendid example of the difficulty of synchronizing such terrestrial accumulations with the marine tertiary deposits named eocene, miocene, and pliocene.

not entirely wanting, are, at all events, feebly exhibited in Northern Europe. But passing from the survey of these valuable exceptional cases, which have been left for our instruction, I will now point out a few examples illustrative of the manner in which several consecutive Alpine formations have been first convoluted, then often inverted, and finally snapped asunder by enormous faults. To treat such a subject in the manner it deserves would require much more detailed knowledge than I possess, and the present notice must, therefore, only be viewed as affording data to assist in explaining the origin and progress of such great mutations.

Let the geological features of any one region of the Alps be appealed to, and it will be seen, that whatever be the major axis of the crystalline mass* in its centre, such also is the prevailing direction of all the sedimentary deposits which lie on either side. Thus in the Eastern Alps, we see two principal ellipsoidal ranges of granite, the one extending from the Iffiger Spitze above Meran to the environs of Brunnecken†, the other of nearly equal extent in the high region near the sources of the Mur, and extending along the left bank of that river to form the nucleus of the Noritian Alps. These ellipses, trending from W.S.W. to E.N.E., mark distinctly the major axis of the Eastern Alps; whilst to the south of Vienna the prolongation of this axis is indicated in the nucleus of the Leitha Gebirge‡. Now this direction from W.S.W. to E.N.E. is likewise that which has been impressed on all the sedimentary masses of these Eastern Alps, of transition, secondary or tertiary age, whether they be successively examined northwards to the valley of the Danube or southwards to the plains of Venice. Minor parallel ellipsoids of crystalline rock, indeed, appear in the Venetian Alps both at Recoaro and its neighbourhood and in the Cima d'Asti, which, whether they be mica schists or granitic rocks, have the same relations to the enveloping younger sedimentary deposits. Such also are the major axes of the great masses of crystalline rocks which occupy the central tracts of the Tyrol, the chief part of the Alps of Lombardy, and the nuclei of the Swiss Alps, and such also is the dominant strike of all the associated sedimentary deposits in these regions.

To the west of the longitude of Berne the chain assumes more of the north and south direction, and there again the sedimentary rocks, to a great degree metamorphosed, run parallel to the axes of the rude ellipses of Mont Cervin and Mont Blanc and their prolongations. And here it is to be remarked, that as we follow the chain from N.E. to S.W. we pass from the clearest types of the sedimentary rocks, and at length in the Savoy Alps are immersed in the highly altered mountains of secondary limestone before described. I am unable to define the manner in which the chief axes of these moun-

* The word 'crystalline mass' is meant to include granite, gneiss, mica schist, marble, &c., and in short all rocks, whether formed by eruption or by metamorphism of pre-existing deposits, which are now in a crystalline condition.

† M. von Buch specially called my attention to this ellipsoid of granite, around which all the rocks are powerfully metamorphosed (see *ante*, p. 167).

‡ Trans. Geol. Soc. vol. iii. p. 303; and Map, pl. 35.

tains trend in the Maritime Alps, where it would, however, almost seem that they bend round so as to be confluent with the Apennines and envelope the great depression of Piedmont and Lombardy; thus describing a grand sweep, or in other words an outward semicircular line, of which the Monferrato near Turin is the last external fold. It is enough for my present purpose, to show, that whatever be the direction of the chief crystalline axis of any one region in these mountains, such is the dominant strike of the flanking deposits. Now, whether such axes are marked by the protrusion of granite, syenite, or any other so-called eruptive rock, or are simply occupied by strata which have been metamorphosed, it is manifest that some powerful energy has been exerted throughout and along them, which action has so affected all the sedimentary deposits on their sides, as to produce a parallelism to the central axes, both in anticlinal and synclinal folds and in deep longitudinal fissures. If the valuable detailed maps preparing by M. Studer were published, this fact would be seen as respects Switzerland, and a glance at the admirable map of France of De Beaumont and Dufrénoy amply explains my meaning in regard to that highly dislocated portion of the chain which extends southwestwards from the region around Mont Blanc. For Piedmont and Savoy the reader is referred to the good illustrations of Sismonda, not yet, however, brought into one view.

In treating of the whole chain it must be admitted, that the Swiss and Savoy Alps have been most agitated; and it is in these most convulsed tracts that we may perhaps best learn what has been the nature of the movements of the strata and the order in which they have followed each other. In parts it is clear, that from the jurassic rocks to the "flysch" inclusive, there has been a continuous series of submarine deposits (see figs. 3, 4, 12, 14, and the group of sections of Hoher Sentis, Plate VII.). Many deep denudations, indeed, expose the whole of this series in lofty mountains on either side of deep valleys, each formation in conformable apposition. The most remarkable fact in this collocation is, that all these strata from the eocene downwards, have been thrown into undulations both rapid and gentle, and sometimes have been so contorted as to produce absolute inversions. I believe that such flexures were among the earliest of the great physical changes impressed upon these submarine strata, which, at the time when they were so bent, may I conceive have been of no greater solidity and compactness than many of the soft deposits which now constitute the crust of the earth in Russia* and other countries, where the processes of induration and crystallization have not been carried out. It seems to me, that however we may attempt to detect the power which produced these folds and contortions, we must admit that all the strata so folded together, had been accumulated the one over the other under the sea (often continuously), and could only have been slightly solidified before the operation commenced by which they all partook of common and conformable movements of undulation.

In no part of the Alps, which I have examined, are the curvatures

* See Russia and the Ural Mountains, vol. i. *passim*.

of the calcareous formations better exhibited than in the Altorf branch of the Lake of the four cantons,—that noble transverse fissure which penetrates so far into the heart of the chain (see fig. 12, p. 195). On the mountain slopes (often vertical precipices) on both sides of this deep cleft, various formations from the Oxfordian or Upper Jura (*o*), near the water's edge (Tell's Chapel), through the lower and upper neocomian, greensand, gault, and sewer-kalk, or equivalent of the chalk, up to the nummulitic and flysch rocks, are all seen to be twisted, and often conformably to each other, in numerous flexures, which increase in rapidity and intensity (in the Achsenberg for example) as you approach the centre of metamorphism (or towards St. Gothard), and decrease as you recede from it. In other words, the folds open out into broader and less complicated sweeps in proceeding from the north slope of St. Gothard as a centre to the flanks of the chain, where they expand into the canton of Lucerne. Some of these extraordinary appearances near Altorf and in the escarpments of the adjacent lake have been figured in two coloured diagrams by Dr. Lusser †. Faithfully delineating what he saw, and judging from the order of superposition, that author concluded, that rocks with green earth and nummulites were repeated several times over in the series, and that these fossils existed in strata (occasionally crystalline) of considerable antiquity, as well as in younger beds. The effort which Dr. Lusser made to classify the rocks of this disturbed tract by mineral characters and *apparent* order of superposition has, I need scarcely say, proved invalid; for as soon as you extricate the nummulitic zone from the labyrinth in which it is involved in the Achsenberg near Altorf (see fig. 12, p. 195 *ante*) and follow it out towards the N.N.W., it is seen to fold regularly over upon the surface of the cretaceous rocks, first in the sharp and partially broken synclinal of Syssikon, then in the dome or anticlinal of the mountain above Brunnen, and next in the broader synclinal of the valley of the Muotta.

The precipitous faces of rock on the sides of the lake of Altorf are indeed most instructive, in showing us the intimate connection between the chief axial line, the folds of the strata and the lines of fracture. In one portion of the lake, nearly midway between Brunnen and Fluelen, the centre of the folds of one of the masses appears in the opposite cliffs, and thus marks the general strike of such contortions to be parallel to the axis, or E.N.E. and W.S.W.; whilst a line of fracture equally visible on both sides of the transverse fissure is also parallel to the same (see *, fig. 12). In short, the order of operations seems undoubtedly to have been, ***first contortion and then fracture***; the nuclei, or inner rolls of the folds, and the lines of dislo-

† Nachtragliche Bemerkungen zu der geognostischen Forschung und Darstellung der Alpen, vom St. Gothard bis am Zuger-See. Swiss Transactions, vol. i. p. 44. Although he does not appear to have noticed organic remains in these mountains, De Saussure has described some of their remarkable flexures and breaks. He speaks of the calcareous strata of the Achsenberg as exhibiting the form of the letter S several times repeated with fractures, and reminds us that Vallisnieri in his 'Origine delle Fontane' had remarked upon these grotesque outlines. He also mentions a great bend in the form of a C from which the strata extend horizontally below.—Voy. dans les Alpes, vol. iv. § ix. 1933 *et seq.* (see my fig. 13).

cation being parallel to each other and to the great axis of the chain. In tracing some of these folds, we see so clearly that an upper stratum has been twisted under one of greater antiquity (and which underlies it at a little distance), that we thus learn a lesson on the small scale which we may endeavour to apply to extensive masses; whilst some of the fractures are observed to have taken place along those portions of the flexure which have least resisted. As my chief attention was specially given to the cretaceous and supracretaceous rocks and their relations, I seldom endeavoured to grapple with the accumulation of obscurities, including metamorphism, which present themselves as the observer approaches the watershed of the chain; it having been sufficient for my purpose to note how the strata in question were uncoiled as they rolled over in great undulations from the centre to the flank. In continuation therefore of a description of the transverse section which passes from Altorf to the N.N.W. (fig. 12), I must in justice say, that, as far as mere outline goes, the undulations seem to conform to the wave-like progression so ably laid down by Professors Henry Rogers and W. Rogers in their map and sections of the Appalachian chain. In other words, the steeper sides of the anticlinal are the most remote from the axis, whilst the longer and less inclined face of each anticlinal faces the chain. This is observed first at Syssikon, and next it is remarkably well seen near the mouth of the Muotta-thal, the structure of which has been described. The nummulitic and cretaceous rocks on the south side of this valley are highly inclined and almost vertical, whilst on the north side they slope at the gentle angle of 20° to 25°. In the next grand curvature of these masses, or towards the Rigi, a tremendous dislocation has occurred*, by which, in fact, *the younger portion* of the nagelflue and molasse of pliocene age is brought with an inverted dip against the escarpment of the lower cretaceous rocks in the manner described in the above diagram. Doubtless this last is a fault of many thousand feet. The axis of the molasse external to the chain, runs parallel to it, as before mentioned, in the environs of Lucerne. Throughout an intermediate distance of several miles there is a development of all those massive and inclined strata of conglomerate and sandstone which form the Rigi. The youngest bed, therefore, of all that vast accumulation is thus brought into contact with, and apparently dips under, lower cretaceous rocks; and as the beds of pebble and sandstone must once have overlapped the cretaceous masses, nummulite rocks and flysch *out of whose materials they have been formed*, the fault must indeed be as enormous as the inversion is astounding.

This grand solution of continuity between the cretaceous rocks with their overlying companions, the nummulites and flysch on the one hand, and the molasse and nagelflue on the other, is the most striking dislocation in Switzerland. The line now mentioned trends from the flanks of Mont Pilatus and passes by the south side of the Rigi, to

* There are other minor folds, and probably dislocations, which I did not follow out, in the masses of cretaceous and neocomian limestone between the Muotta-thal and the Rigi. The dome shaped arrangement of the sewer-kalk at Sewen indicates that this must be the case (see p. 193).

the east bank of the lake of Lowerz, where it marks the junction between the lofty cretaceous peaks of the Mythen above Schwyz, and the supracretaceous rocks of the Hacken pass and Lowerz. But here dislocated masses of flysch and fractured nummulite rocks are intercalated between the cretaceous escarpment under which they seem to dip, and those great sloping masses of conglomerate, which, constituting the Rossberg so celebrated for its landslip, appear in their turn to underlie the nummulite zone. This inverted position is again well displayed as you follow the same masses towards Einsiedeln, where the nagelflue overlying the middle and lower molasse is in distinct apposition to an escarpment of nummulite limestone, which dips rapidly away under mountains of flysch that are also thrown off to the S.S.E., or *towards the axis of the Alps* (fig. 13). This phænomenon is common along the northern flanks of the chain. It is, in fact, that prevalent feature throughout the external zone of the Eastern Alps on which Professor Sedgwick and myself insisted; but at that time we had not an adequate conception of the intensity of these movements, by which, on lines parallel to each other, the oldest portion of each group has often been thrown up on the external or younger side of the Alps, with its last-formed member let down as it were, so as to be in contact with the oldest rock in the tract, and with all the appearance of passing under it!

The distinctions between regular succession and discordance are admirably displayed around the Grünten mountain and between it and the higher Alps; for after an exhibition of perfect symmetry (figs. 17 and 18), we find the flysch truncated (fig. 19) against a wall of cretaceous rocks. We pass through that wall by the gorge of the Hirsch-sprung, and again we have undulations and slopes occupied by upper members of the series which are entirely lost on the steep side of the anticlinal. Again, at the outer or northernmost escarpment of the Grünten (fig. 18), we have the same tremendous fault as that before spoken of along the Rigi and Rossberg, showing the nagelflue and molasse in juxtaposition with the lower neocomian. In this last case, however, the molasse is rather thrown off to dip away from the secondary rocks; but along the same line of fault, and immediately to the west of the river Iller, all the mountains of nagelflue again appear to plunge directly under the zone of flysch. They there mark the grand outer line of disseverment between the molasse and all pre-existing strata, which trending from near Immenstadt in the Allgau, passes by Dörnbirn and Haslach* south of Bregenz. This same line of fracture is again magnificently displayed in the canton Appenzell, along the precipitous north-western face of the Hoher Sentis. There, the upper portions of the enormous masses of molasse and nagelflue, dipping away to the S.E. from the St. Gallen axis before-mentioned, occupy mountain pasture tracts†, whose

* At Haslach near Dörnbirn, on the right bank of the Rhine, the nummulite rock is so collocated, that any one ignorant of fossils would really believe that it passed under the limestones of the Stauffen, composed of lower and upper neocomian rocks.

† Handwyler Hohe, Kronberg, Petersalp, &c., these conglomerates range into the Speer mountain, and thence to Wesen.

sharp and peaked ridges have in some places the high inclination of 65° to 70°.

When viewed longitudinally in the little valley of Weissbad, nothing can be more striking than the aspect of these bold tertiary peaks on the one hand, and the massive cretaceous precipices on the other, under which they seem to dip*. Examination, however, shows an enormous void between these two classes of rocks. The upland valley is indeed encumbered with much detritus, as is frequently the case along such lines of fault, and for the most part fragments only of dismembered flysch, with a rare specimen of nummulite limestone, are to be seen as memorials of the vast destruction of intervening rocks which has occurred. In one spot, however, at a little cascade under the Thurm, one of the buttresses of the Sentis, I detected a portion of the flysch, which is fairly bent under the cretaceous masses of the mountain, which I believe to represent the sewer-kalk or chalk; for in the heights above this cascade, Prof. Brunner and myself reached, after some peril and labour, a zone of secondary green sandstone. M. Escher has, indeed, shown that the chief culminating masses are sewer-kalk or chalk based on greensand and neocomian. That author pointed out to me, that the Sentis group is not merely a double or triple chain, but is made up of six lines of ridges, in which the greensand and chalk are repeated with supracretaceous troughs. He has drawn for me the diagrams in the annexed plate (Pl. VII.), which are the result of his long and patient examination of this remarkable tract. These transverse sections, made at short intervals from each other, will explain better than pages of description, how the apparent alternation of formations, whose denuded edges crop out to the surface, is due to folds, the axes of which, though occasionally vertical, are usually oblique or inverted towards the high chain of the Alps, and thus often present their chief escarpment to the hills of younger tertiary conglomerate. By this arrangement, nummulitic eocene rocks (*f*, *g*) dip for the most part under strata of anterior age; and whilst, on the S.E. face of the mountain, they plunge towards the Alps (there regularly overlying the chalk and greensand), on the north-western side they are truncated between the molasse (*m*) on the one hand and the cretaceous rocks (*a*, *b*, *c*, *d*) on the other, but usually dipping under the latter.

Another most instructive section, and parallel to the above, is that which proceeds from the molasse and nagelflue, the mountain called the Speer (fig. 14, p. 200) on the N.N.W. across an inverted, inclined axis, which clearly exposes nummulite rocks and sewer-kalk on either side of a nucleus of neocomian limestone; whilst by another fold the whole series up to the flysch is displayed in a lofty basin, where the inoceramus limestone rises rapidly into the lofty mountain Lyskamm, from which, after some undulations, we see a regular descending order through the whole cretaceous rocks and the jurassic system of this region, as displayed in the cliffs on the north side of the lake of Wallenstadt.

If each Alpine region be examined in detail, and its geological fea-

* See fig. 14, p. 200, and plate of M. Escher's Sections, pl. vii. In the latter the six ridges alluded to are numbered i. to vi. I have pointed out the transition bed (*e*) and have distinguished the eocene from the cretaceous.—R. I. M.

tures laid down on maps in the manner in which L. von Buch, Prof. Studer and M. Escher de Linth are working them out, it will be seen, that although their major axes have a strike from E.N.E. to W.S.W., there are numberless local deviations, and sometimes to a very considerable extent. In fact, it is in the very nature of the formations which clasp round such ellipsoids as those before spoken of, that they should present local aberrations from any one chief line. Such divarications occur in the masses which surround the great ellipsoid of the Grisons and the canton Glarus; for although the major axis of that tract proceeds from E.N.E. to W.S.W., the strata where they conform in outline to the *ends* of the ellipse, depart considerably from the normal direction. I examined this phænomenon on the north-eastern portion of the external zone of this great ellipsoid in the company of M. Escher, viz. in the environs of the lake of Wallenstadt; and as a map of this tract was coloured for me on the spot by my companion, I have exhibited it to the Geological Society, to illustrate the phænomenon under consideration. To attempt to describe this tract in words would be in vain, and I therefore content myself with saying, that this map shows, that whilst the chief anticlinal and synclinal lines conform to the general axis of the chain, the rock masses of various ages, from the jurassic to the nummulitic rocks and flysch inclusive (which in the chief ridge of Sentis and along its outer face strike E.N.E.), are bent round to the S.E. and S. at the east end of the lake of Wallenstadt and in the valley of the Rhine near Sargans. In this short space the rocks, therefore, become strikingly divergent from the major axis, or in other words, *they fold round the extremity of the ellipsoid*. I must leave others to expatiate on the phænomenon, which will be the better understood when M. Studer shall have developed all his views, and when it may be ascertained, that the massive ellipsoids of Mont Blanc, the Finsteraarhorn, the St. Gothard, La Selvretta, &c. have been acted upon by subterranean forces peculiar to each, and yet all partaking of one common line of direction.

It is worthy of remark, that just as the metamorphism of the rocks is greater as we approach the centre of the chain, so do the sedimentary masses the more arrange themselves on the surface, as if their external configuration were intimately connected with some *grand crystalline change*. On the other hand, as we extend our researches to the outer zones of the chain, we pass over *numerous folds and breaks*, all of which are evidently referable to pure mechanical agency. Thus, on the N.W. face of the synclinal valley of Wildhaus, we meet with the system of flexures in the Hoher Sentis already alluded to (Pl. VII.), whereby the neocomian, greensand and chalk are repeated on lines trending due N.E. and S.W., and forming the ridges and troughs of that remarkable group, slightly divergent from parallelism to the true axis. In alluding to the synclinal troughs which run parallel to the major axes of the Alps, it is to be observed, that in one tract the same trough will be found unbroken, which when followed in its direction, shows different degrees of rupture. One of the troughs before alluded to in the promontory of Bürgen (figs. 9 & 10, p. 192), on the west side of the Lake of the four cantons, or in other words, the prolongation of the great synclinal occupied by the Alpnach branch of the Lake of the four cantons, is seen to constitute a good massive synclinal-formed hill, the promontory of Bürgen, in which the nummulite and flysch rocks are troughed on neocomian and cretaceous limestones; but if followed to the opposite side of the lake to Viznau, viz. in the same direction (in the space of two or three miles), that which was in a synclinal form has become the scene of that grand fault or rupture by which the upper nagelflue plunges against and apparently under the neocomian, almost to the exclusion of the nummulite and flysch rocks, fragments of which only appear (fig. 12, p. 195). Following on this line to the N.E., across the lake of Lowerz, the representatives of the nummulite rocks and flysch are intercalated, though in a highly broken condition, between the molasse and nagelflue of the Rossberg and the cretaceous rocks of the Mythen; whilst, still further to the N.E., these nummulitic rocks, so squeezed up on the flanks of the Mythen, expand into the tracts south of Einsiedeln, where I have mentioned them as having an inverted dip, or towards the axis of the chain (fig. 13, p. 197). Thus, that which is an overlap in one portion of the sides of a synclinal, and whereby an enormous transposition or slide of the masses has occurred, often occasioning the absolute destruction of copious formations along the line of fracture, on another part of the same line is, as far as external appearances go, a complete overthrow, in which the older rocks are superposed to the younger.

As the same physical relations of the rocks, whether in anticlinal or synclinal forms, are seldom persistent for more than a few leagues, and rarely in absolutely right lines, so but few of the longitudinal faults are continued for great distances without interruption or change in their conditions; and although some of them pass across transverse valleys without much deviation from their strike, it is not unfrequent to see a considerable lateral displacement, or as it were a movement "en échellon," in masses occupying the opposite sides of any broad transverse valley. In crossing the valley of the Rhine, for example, near its "débouché" into the lake of Constance at Bregenz, in the direction or continuation of the synclinal flysch valley of Wildhaus, we find a large outlier of cretaceous rock at Eschen on the right bank, which is in fact an anticlinal of neocomian, flanked by sewer-kalk or chalk, and trending N.E., whilst the chief trough or synclinal of flysch setting on to the south of Feldkirch, trends decidedly E.N.E. across the Ill*.

The great cretaceous masses of the Hoher Sentis are repeated or continued, it is true, in a general way in the mountains of the Hohe Kugel and the Stauffen (the insulated hill of Kamor in the valley of the Rhine serving as a link between the opposite promontories); but there the nummulite limestone, instead of being thrown off the cretaceous rocks, as in the Fähnern mountain on the left bank, as before explained (fig. 15), is abruptly collocated with an inverted dip (fig. 16) against a grand neocomian escarpment; whilst between that junction

* See Würl's Map of Switzerland, which is recommended strongly to all geographers and geologists.

and the molasse, the beds of flysch are exposed vertically in the synclinal of the valley of Oberdorf. This Dörnbirn zone of nummulite and flysch rocks is therefore the third parallel trough on the right bank of the Rhine, reckoning from the higher and central Alps, just as the zone of the same rocks in the Fähnern, which is almost lost in the fault of the Weissbad valley, is the third repetition of such formations on the left bank of that river, reckoning from the copious mass of it in the high mountains of Glarus, which extends from the heights of Harstock across the valley of the Sernft by Elm and Engi to the baths of Pfeffers and the environs of Sargans on the N.E. In one portion of the outermost of these folds, or that of the "Fähnern mountain," we have seen how symmetrically the nummulite rock and flysch overlie the cretaceous rocks; whilst on the same line on the north, a flank of the Hoher Sentis, a few miles distant only, the whole formation is obliterated by the great fault. In the second or intervening zone of Wildhaus, between the Sentis and the Kurfursten mountains, the nummulite rocks and flysch are regularly troughed upon cretaceous rocks. In the inner parallel, however, or that nearest the axis of the chain, the phænomena of inversion "en masse" so exceed in grandeur anything of which I could have formed an idea, that I must direct attention to it, particularly as I had the advantage of travelling over a lofty portion of the inverted tract in company with M. Escher, who has the exclusive merit of having worked out the data.

Grand inversion of masses in the Canton Glarus.—Ascending the valley of the Sernft by Engi to Elm, M. Escher and myself thence traversed the Martin's-loch pass, about 8000 feet above the sea; and in this ridge, which separates the canton Glarus from the Grisons, I saw the rocks which I now describe (fig. 28). The lowest

Fig. 28.

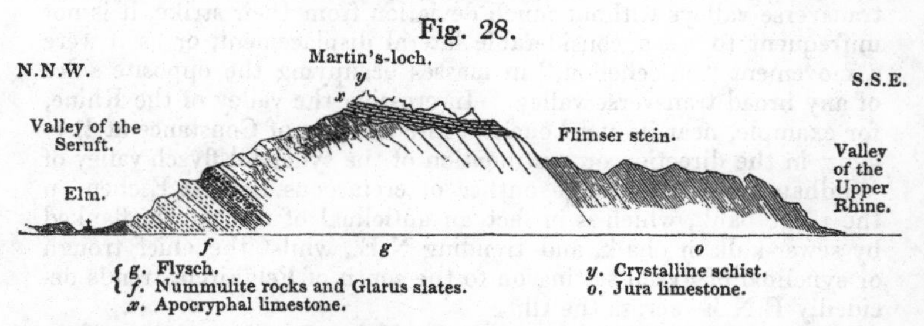

{ g. Flysch.
{ f. Nummulite rocks and Glarus slates.
x. Apocryphal limestone.
y. Crystalline schist.
o. Jura limestone.

visible strata are schists and Glarus slates, the continuation of those containing fishes, and with them sandy calcareous grits and limestones with green earth and nummulites (*f*). These bands plunge directly into and under the mountain, or to the S.S.E., and are overlaid by a very quartzose variety of the flysch (*g*) which seemed to me to be a partially altered rock. On the sloping surface of these grits we detected a few loose fragments of limestone with Inoceramus and Belemnites which seemed to have fallen from some adjacent summit. The flysch, however, continues to be the chief rock of the mountain until you reach the depression in the high ridge where the track passes into the Grisons, and the crest is there so narrow and elevated, that we positively sat upon the peaks of flysch with one leg in the Grisons and the other in Glarus. Widening to about 100 feet or more to the south-west of the mountain-path, this flysch is then directly surmounted by a mass of hard grey subcrystalline limestone (x) (about 150 feet thick), which is perforated by a natural tunnel or hole*, and hence the name of Martin's-loch. This limestone is, as far as my eye could discern (and it commanded several miles both to the east and west), continuously superposed to the flysch in varying and irregular thicknesses, and more or less in a tabular position, over a great area, including the peaks of Hanstock, Linterberg, and Kärpfstock. M. Escher had, indeed, sedulously followed the range, and had found in it jurassic ammonites near the Kärpfstock. Now, this limestone is in its turn distinctly overlaid by a zone of talc and mica schist (y), in parts having quite the aspect of a primary rock. This uppermost rock, according to M. Escher, is an integral part or continuation of the Sernft conglomerates and schists which are seen in the adjacent vale of Wallenstadt to lie beneath the whole secondary series. Before I made this section, I had supposed that the younger and nummulitic deposits might be simply plastered up on the sides of the older rocks, and not really pass under them. But the examination of the lofty and narrow ledge I had traversed checked such hypotheses, for on both sides of it I witnessed the same relations. Again, I tried to imagine, that without any inversion of the strata, metamorphism had here seized upon all the upper strata to the exclusion of the lower; but this speculation was equally fruitless; for, independently of the proof obtained by M. Escher, that the overlying limestone contained ammonites, that rock is quite unconformable to the flysch on the edges of which it reposes irregularly. I was also well assured from pretty extensive observation, that no such rock existed in any part of the supracretaceous series. In descending from the summit-ridge into the valley of the Vorder Rhein in the Grisons, I had, indeed, another and an independent proof, that the rocks *underlying* the solid limestone, with its cover of talc schist, were really of supracretaceous age, for we found in them both nummulites and the same teeth of fishes which characterize the flysch in many other tracts. At this point the fossiliferous "flysch" beneath seemed to be quite uncon-

* M. Escher informs me that the superposition of the jurassic rocks to the nummulitic extends to the Rosenlair mountain, in the canton of Bern, and to the Grisons. In the canton Glarus he has found the same relations to range from Martin's-loch to the Panix pass, where crystalline schists, equally resting on the limestone in question and on nummulite rocks, are further surmounted by a limestone with Pentacrinites, and resembling the modified inferior oolite and lias of these regions. If ever, therefore, it should be attempted to explain away the anomaly of Martin's-loch (as M. Escher well observes in a letter to me), by supposing that its enigmatical limestone and overlying chlorite schists are mere modifications of true overlying "flysch" on Jura limestone, still the superposition of the pentacrinite limestone to the whole of the series is inexplicable except on the supposition of a complete overthrow "en masse."

formable to the overlying limestones and talc schists. In proceeding, however, to the opening from the glacier of Segnes, where the waters issuing from a small lake or tarn, rush through crevices in the secondary (Oxfordian) limestone, the very same masses of flysch seem to dip under that limestone, which in its extension occupies the striking ridges called the Flimser Stein, on the left bank of the Vorder Rhein. Yet, these very same masses of jurassic limestone, so inverted in the tract described, when followed to the heights south of Pfeffers Baden, are found to plunge under the whole of the massive limestones of neocomian and cretaceous age, and finally to be surmounted by nummulite rocks and those grand masses of flysch from which the mineral waters issue; and thus, in proceeding towards the lake of Wallenstadt, or towards the flank of the chain, all is symmetrical and each rock resumes its normal position. Whether therefore I examined the pass of Martin's-loch and its respective sides, and looked at its absolute sections, or cast my eye to a distance over the terraces of limestone surmounting flysch and nummulite rocks as seen from its lofty summit, I was convinced that M. Escher was correct in his delineation and mapping of the ground, although he ingenuously urged me to try in every way to detect some error in his views, so fully was he aware of the monstrosity of the apparent inversion*.

I dare not pretend to offer an explanation of the "modus operandi" by which such a marvellous mutation of order has been produced over so vast an area. I had indeed previously witnessed every possible contortion on a minor scale, and I might think it only necessary to amplify the *measure* of such movements. But it became necessary to admit, that the strata had been inverted, not by frequent folds, as on the sides of the lake of Altorf or in the Hoher Sentis, but in one enormous overthrow; so that over the wide horizontal area above-mentioned, the uppermost strata which might have been lying in troughs or depressions due to some grand early plication, were covered by the lateral extrusion over them of older and more crystalline masses; the latter having been forced from their central position by a movement operating from centre to flanks, or in other words, from the axial line of disturbance towards the sides of the chain. One inference, indeed, seemed certain, that if the masses have been thus inverted, there must have since occurred enormous denudations to leave the older limestone and talc schist merely as the narrow cappings which they form on the summits of the ridges in the manner represented in fig. 28. The grandeur of this phænomenon may to some extent be imagined by consulting that section; but a true conception of it can be alone formed by climbing over the ridges in which the facts are laid bare, in one of the most pictorial regions of the Alps. Not the least extraordinary feature of the phænomenon is its apparent uniformity, simplicity and grandeur, and the absence throughout the tract of those mechanical plications, which, as we remove our ob-

* The map and sections in which this startling phænomenon is recorded, are published in the work on the statistics of the Canton Glarus, by Professor Heer, previously referred to.

servations from the supposed centre of disturbance, become so manifest in the outer group of the Hoher Sentis in the same longitude. I am aware that there is a great difficulty in accounting for the lateral folds and plications of the Alpine strata, from the supposed absence of adequate central masses of erupted matter to dislodge, roll over or compress into smaller horizontal areas, strata which must once have been regularly extended in sheets. But might not the formation of great central crystalline ellipsoids, whether eruptive or metamorphic, serve in some measure to account for this? May not these ellipsoids, in being transformed and amplified, have operated as great centres of mechanical force? And with our knowledge of the position here and there of very considerable masses of true granite, may not much of that rock have acted without being visible, and may not large masses of it be hidden under unfathomable glaciers?

But leaving this enigma, let us return to the consideration of the lateral folds to which the strata of these mountains have been mechanically subjected. In them we learn not to be sceptical concerning the truth of many sections in the Alps, such in particular as those of M. Hugi, which represent rapid repetitions of lias and different jurassic formations in parallel sheets; for we need only suppose the superficial portion of narrow undulations removed by a powerful denudation, and many of the phænomena he represents would be at once realized.

I am happy to bring forward these few data at the present moment, when Professor H. Rogers, one of the authors of the undulatory or earthquake theory as applied to mountains, is in England, and when he has taken the trouble to point out to me how *some* of my facts may, as he thinks, be explained on his principles of illustration. Putting aside his theory, we have only, indeed, to look at the elaborate map of the Appalachian chain, by his brother and himself, and witness the numerous ellipses into which the palæozoic masses have there been turned, and scan the sections of these authors, based on positive data and outcrops of mineral masses worked for use, in order to comprehend how the enormous faults and slides have there occurred just where the strata have been most bent and inverted in reference to the centre of disturbance. Thus, the comparatively low chain of North America may throw light on some of the most complicated problems of our science, which could scarcely ever have been satisfactorily worked out amid the confusion of the Central Alps, such large portions of them being inaccessible to man and covered with eternal snow.

The inversion or the dipping of the strata towards the centre of a chain, so as to place the older over the younger deposits, has been a subject of wonder, and has hitherto been considered scarcely explicable upon any satisfactory hypothesis. In viewing the Ural mountains, where the same phænomenon is copiously displayed, I was disposed to account for such apparent inversion, by supposing that the broken ends of the strata had fallen into abysses or cavities produced by the extravasation of the enormous masses of igneously formed rock, which are there seen at hand as if ready to explain the facts.

But this explanation is totally inapplicable to the Appalachians. It is almost impossible also to apply this reasoning to the Alps, from the absence of masses of erupted matter adequate to account for the phænomenon by displacement. But, however we may theorize, it must be admitted, that in nearly all the alpine folds to which my attention was directed, the longer leg of each anticlinal slopes towards the centre of the chain, whilst the steeper talus or shorter leg of the flexure is away from it (see figs. 12 & 14, and Plate VII.). Besides the occurrence of this phænomenon, which is the basis of the theory of Professor Rogers, the Alps seem further to exhibit, as far as I know them, the same longitudinal faults as the Appalachians, whereby fractures having occurred either on the most bent portion, or the steep side of the anticlinal or synclinal folds, the result has been (explain it how we may) the lateral overlapping of the older rocks upon the younger. In saying that I am not prepared to subscribe to the earthquake theory, I have to thank Professor H. Rogers for having drawn diagrams to explain two of the most frequent cases of such overlap and inversion, as they occur indeed in my own sections, showing how the axes of each trough or ridge were first forced into oblique positions, followed by the fractures in question, and then by the transgressive sliding of older over newer deposits by lateral pressure. The following is his explanation.

" I have endeavoured (says Professor Rogers) in the annexed diagrams to illustrate two very common kinds of faults or dislocations occurring in regions of closely-compressed or inverted flexures. In one case (fig. 29. Nos. 1, 2 & 3) the fracture coincides, or very nearly so, with the anticlinal axis plane and the plane which cuts the two branches of the anticlinal flexure at the same angle; the other instance (fig. 30. Nos. 1, 2, 3) is where the dislocation is in the synclinal axis plane. The displacements here shown are both of them upcasts along the inclined plane of the fault. In all oblique compressed flexures, this plane of the fault *dips of necessity towards a more disturbed side of the district*. The effect of both of these classes of fracture is to bring an older set of strata superimposed in approximate parallelism of dip upon a newer series, but with opposite conditions, the anticlinal fracture inverting the beds on the side below or beyond the fault, while the synclinal fracture inverts those on the upper or nearer side: I think it will be found that the first phasis is by far the most common in the Alps. The greater part of the dislocations of the Appalachian chain are certainly of this character, the fracture being either in the anticlinal plane or a little beyond the axis, in the short inverted leg of the flexure. Most of the cases of inversion in the Alps which your interesting sections display, and to which you have kindly drawn my attention, are, I think, *simply instances of dislocation along the anticlinal planes of inverted or closely-compressed oblique flexures*. A few, however, appear to have resulted from faults along the synclinal planes. I have not here exhibited the other less usual forms of dislocation, or treated of the cases where the displacements are downthrows and not upthrows along the inclined plane of the fault."

This ingenious explanation of Professor H. Rogers may, it appears to me, be very well applied to those examples in the Alps where, as assumed by him in his diagrams, the strata of different ages have originally been continuously and conformably superposed. Such,

Fig. 29.

Fracture through the anticlinal axis-plane of an inverted flexure (the elevated mountains are to the right hand).

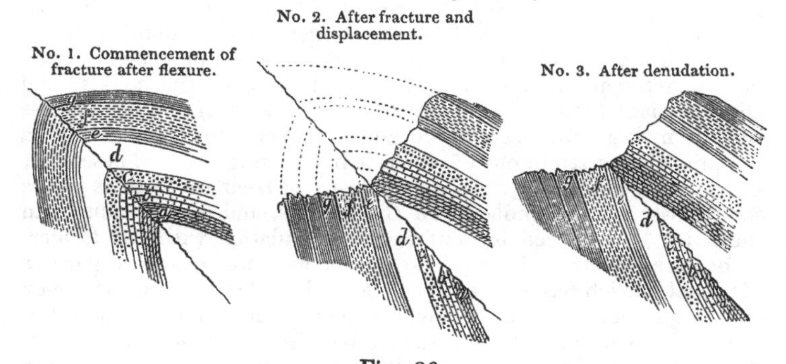

Fig. 30.

Fracture through the synclinal axis-plane of an inverted flexure.

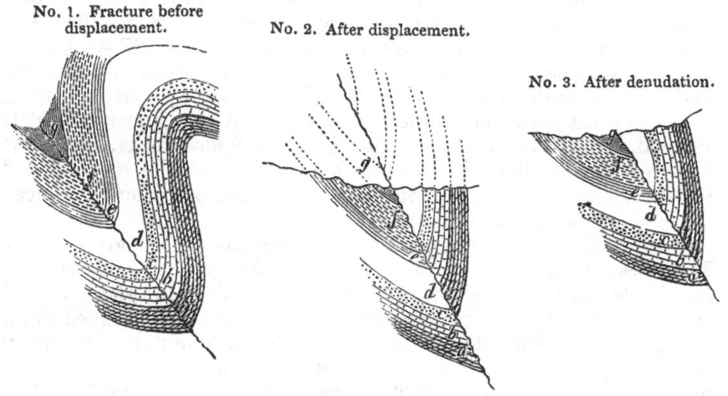

for example, may have been the case in all those tracts where the cretaceous rocks were formerly surmounted by nummulite limestone and flysch, and where, after having been thrown into inverted anticlinals and synclinals, they were afterwards fractured and denuded in the manner described. Of this class of faults the figures 16 and 19 may be cited as very *probably* answering to the law of displacement observed in the United States*. In the first of these, near Dörnbirn,

* I say *probably*, because in the Alps the subterranean course of faults has not been ascertained by mining operations as in the United States, and examination is usually much impeded by vast quantities of detritus.

the nummulitic strata (*f*) which are inclined at an angle of about 45°, may really pass under the truncated edges of the lower neocomian limestone (*a*) in the manner represented. In fig. 19 the beds of flysch of the Bolghen may be similarly overlapped by the neocomian limestones of the Schwartzenberg in a similar manner, though probably the angle of the line of fault is different from that at Dörnbirn. In both these cases, as in many others throughout the Alps, where the pressure has been so exerted from the centre towards the flanks of the chain as to invert the axes of a series of formations *originally conformable*, the law or rule of Professor Rogers may be found to apply. On the other hand, a different method of reasoning may be necessary, in reference to the great Swiss fault between the molasse and all pre-existing rocks (see figs. 12, 13, 14, 17, &c.); for in no case was the molasse and nagelflue originally superposed conformably on the pre-existing strata out of which it has been formed. These older strata must, in fact, have undergone many of their contortions before the molasse was accumulated on their flanks, and in no instance can the latter be observed in conformable undulation with the former. In no case (as far as I know) do the molasse and nagelflue partake of any fold which has affected the older rocks. On the contrary, they are always abruptly truncated against such older strata, and more frequently with an inverted dip than otherwise. It is indeed manifest from the composition of the nagelflue, that when it was formed, the secondary rocks of the Alps, and even the nummulite rock and flysch, were hard solid masses; in fact, just in the lithological state we now find them. Again, we cannot look at the sections on the flanks of the Rigi, Hoher Sentis, &c. (figs. 12, 13, 14, pp. 195, 197, 200, &c.), where the upper conglomerate of the molasse is forced into inverted and unconformable contact with the older rocks, without perceiving that the great anticlinal of the regenerated and younger strata (*m*) is vertical, and not inverted as in the older and folded rocks. And this fact teaches us, that the forces which upheaved the molasse and nagelflue were independent of those which contorted and produced the earlier fractures in the chain.

But whatever view we may take, the phænomena of the group of the Hoher Sentis present us with very remarkable problems not easily reconcilable (see Pl. VII. p. 243). Whether viewed from the plains around the lake of Constance, or examined in its precipitous and rugged sides, few geologists would doubt that this cluster of mountains had taken up its position relative to the lower country by a great upcast*. Yet nowhere within it has M. Escher been able to detect anything like a centre of upheaval, still less any motive cause of elevation; for the highest summit is not composed of the oldest rock of the chain, viz. the lower neocomian (*a*), but, on the contrary, of the equivalent of the chalk (*d*). Its remarkable features are rapid folds (doubtless accompanied by some considerable faults), by which, in fact, the

* In my section I have hypothetically drawn the line of the great fault between the molasse and the older rocks nearly vertical, but whether it inclines away from the chain, according to the usual form of upcasts, or continues to pass under the older rocks, is not known.

group where most extended is divided, as before said, into no less than six or seven parallel ridges, with intervening troughs; and all this in a very short horizontal distance, wherein nearly all the strata from the lowest neocomian to the flysch above the nummulites are repeated over and over. Now, if these plications with vertical and inclined axes be due to any force which has proceeded from the centre of the Alps, is it not extraordinary that this group of the Hoher Sentis, so far from that centre, should exhibit this extraordinary amount of contortion, and should also in this respect differ so essentially from other parts of the zone of which it is a prolongation? for in following the same band of the calcareous mountains to the south-west, through Switzerland, it is found to be of much simpler contour; presenting seldom, if ever, more than one or two folds and a fault*.

General View of Changes in the Alps.

Whilst the inaccessible forms of large portions of the Alps, their fractures and curvatures, and the enormous piles of rubbish on their slopes, render it difficult to trace accurate sections of them, the general survey of this chain warns us not to infer the *independence of formations* from the unconformable or broken relations of any one tract. Having full confidence in the accuracy of the observations of M. Favre in the region so much examined by De Saussure,—observations the more to be admired as they have been carried out in the very spirit of his illustrious precursor,—let us admit with him that the *terrain à nummulite* near Geneva† and in parts of Savoy reposes on jurassic rocks, or neocomian limestone or greensand, just as it has been observed in the Maritime Alps by Sismonda, and near Chambery by Chamousset. Still, this is only a proof, that in such localities the intermediate cretaceous beds have either not been elaborated, or have been denuded by local causes before the deposition of the nummulite rocks commenced. Such examples of a want of regular sequence cannot be maintained, as M. Favre contends, to be proofs of independence, when set against the examples of superposition and passage into the chalk at Thones in Savoy, in the Appenzell Alps, and in the various parts of Switzerland and Bavaria above cited. The latter must be viewed as the rules of order and succession. Again, judging from the local sections near Samoens and Taninge in

* Professors Studer and Brunner have written to me on the application of the theory of Professor Rogers to the Alps. Though both of them seem to have had in a certain sense a perception of his views, still his explanations of faults appear to me to be distinct from those of any of his precursors or contemporaries. The sections of M. Dumont of the palæozoic strata of the Ardennes and the palæozoic strata around Liege make perhaps the nearest approach to them. Professor Studer had shown, that the undulations of the Jura, as described by Thurman, resulted from the elevation of the Alps (Bull. Soc. Géol. Fr. vol. ix. and Géographie Physique, t. xi. p. 235). But Professor H. Rogers is quite of a different opinion respecting the undulations of the Jura. Judging from their form, *i.e.* with the long slopes towards the French side, and their steep slopes towards the Alps, he infers, on the contrary, that the propelling force came from the Black Forest.

† Bulletin de la Soc. Géol. Fr. 2nd ser. vol. iv. pp. 999–1001.

Savoy, M. Favre believes that the flysch is as independent of the nummulite rock as the latter is of the pre-existing limestone; whilst, if the above-mentioned instance at Thones in the same region was not sufficient to prove the contrary, I have shown by many other examples, that nummulite rocks and flysch constitute one and the same natural group in which no general severance has taken place. I recur to this point, because several continental geologists have insisted on the establishment of the "independence" of formations by an amount of unconformity which in my opinion is simply due to *partial* dislocations and overlappings of the strata. Now, it is quite manifest from the examination of any large region, that movements of the subsoil have occurred in one tract both during and after the accumulation of a deposit, which extending their influence to a certain distance only, have not interfered with the continuous succession of the same deposits in a neighbouring country. Changes of level at various periods, accompanied by contortions and breaks, have often produced those transgressions from which "independence" is assumed; whilst in following out these very masses into other tracts a clear and conformable succession is developed. English geologists need, in truth, no caution on this head, for the phænomenon is well known to them, and it has been recognised on the grandest scale in North America, through the labours of our associates of that continent.

Is there then no formation in the Alps so completely and universally broken off from all other deposits that it is really independent of them all? As to the oldest sedimentary rocks of the chain, it is unquestionably true that some of them (all those at least which are affected by a rude slaty cleavage) are so essentially distinct from the deposits which followed, that we may fairly suppose that they acquired their mutations in an earlier epoch. The most distinct, however, as well as the grandest of the examples of true independence, is that of the molasse and nagelflue of Switzerland, to whose position so many references have been made. As relates to Switzerland and all the northern face of the Alps, these deposits appear to have been so completely dissevered from all pre-existing strata, as to leave a considerable geological *vacuum* between them and the eocene group. It has accordingly been seen, that there is a vast difference in the fossils of the nummulitic group of that chain and those of the succeeding molasse, a difference which induces me to class the latter rather with the older pliocene than with the miocene. But when we turn to the southern flank of the chain, we there find, as I have shown, an apparent conformity from the cretaceous rocks through both eocene and miocene into the pliocene, although the axial line, it is to be recollected, is perfectly parallel to that of Switzerland and Bavaria, where the great hiatus exists. In the Italian case, I believe that another *parallel* elevation, posterior to the great upheaval of the eocene, raised the external fringe of younger tertiary rocks into the hills of Bassano and Asolo. In treating of Italy and the Apennines, I shall, indeed, endeavour to show that those portions of the sections of the tertiary series which are either denuded or imperfectly seen in the tract between the Brenta and the Piave, are taken up and clearly displayed in the Monferrato ridge, and that the Superga exhibits, on the one hand, a downward transition from what has been considered true and pure miocene into nummulitic strata, and upwards, on the other, into the richest subapennine or pliocene marls and sands. The great hiatus on the northern flank of the Alps may represent, perhaps, the upper portion of eocene, and the lower part of what has been termed the miocene age, whilst on the south, evidences have been left, of apparent transitions from one to the other.

The conclusion therefore is, that without quitting the Alps and their immediate flanks, we may argue for or against the independence of several formations, according to the tract we survey. In England the coal is generally conformable to the mountain or carboniferous limestone. But now we know, that what is true in England and the west of Europe, is not so in certain parts of Bohemia and Poland. In these two countries a great dislocation has taken place after the deposition of the mountain or carboniferous limestone with its large Producti, and before the accumulation of the overlying coal-fields; the former being highly inclined together with Devonian and other palæozoic rocks, whilst the latter are horizontal.

Nothing, however, that I have stated must be taken as militating against the indisputable phænomena of dislocations having occurred in one region whilst adjacent countries remained quiescent,—phænomena which often enable us to mark the æras of such disturbances. It is not against such general views of M. É. de Beaumont that I contend, but simply against the abuse of them, in the hands of those who would magnify into too great importance local and partial lines of rupture. At the same time, I cannot doubt that great mutations of outline have taken place at different periods, not only in and along the same chain of mountains on lines parallel to each other, but even at *different periods upon the very same line*. Judging from the analogies in existing nature, such events might well indeed be supposed to happen upon any one line of fissure, where the earth's crust had been once much weakened by rupture. On this point I may revert to proofs, cited by myself in the north-eastern portion of the Silurian region of the British Isles, to show that similarly constituted igneous matter had been successively extruded along the same line of fissure or vent of habitual eruption, at one period mingling and alternating with the Silurian sediments, afterwards throwing them on edge, next affecting carboniferous strata which had been deposited on the edges of the Silurian rocks, and at a subsequent epoch cutting in dykes through the horizontal new red sandstone, thereby isolating a basin of lias*. *Now, all this occurred upon one and the same line at those successive epochs.*

In concluding this portion of the memoir, I must further be excused when I refer to another chapter of the 'Silurian System†' for what I consider to be a true delineation of Alpine phænomena, although on a smaller scale. In the Alps, as in Siluria, we see local divergent strikes, sometimes of considerable extent, amidst rocks of the same

* Silurian System, pp. 294 *et seq.* † See Chapter XLII. p. 572.

age, and parallelism of masses which were formed at different epochs, and in both regions we trace the disturbance and transgression of certain strata in one tract, and their inosculation and quiet transition in another.

In the preceding pages I have endeavoured to present one general view of the successive formations of the Alps, from the earliest periods in which there are traces of life, until that grand rupture occurred, by which the youngest tertiary deposits of the north flank of the chain underwent those tremendous movements, which left them in their highly-inclined and apparently inverted positions. With the exception of the evidences of a very limited terrestrial vegetation afforded by some of the older strata, and again by the lower portion of the eocene or nummulitic group (which can be accounted for by vegetable matter having been drifted into bays or estuaries), nearly all the sedimentary rocks bespeak, through their imbedded organic remains, a continuous accumulation under the sea. Passing by for the present the palæozoic rocks and the trias, as yet only known in the Eastern Alps, and limiting our attention to the Western Alps, we cannot view the grand succession of jurassic, cretaceous and nummulitic formations without perceiving, that although some of them were unquestionably formed in shallow water, even these must have been depressed to very great depths in order to account for the copious and continuous superposition of other and younger marine deposits of vast thickness. In appealing to the series of natural-history records as written on the walls of the Alps, we find that extensive and sometimes entire changes in the animals of these seas took place, even when the beds in which their relics are now entombed appear to have succeeded each other without any general physical fractures or derangements of the then existing surface. In no case is this so remarkable, however, as when the nummulitic or eocene group surmounts by conformable transition the uppermost member of the cretaceous system.

At length, however, a period arrived, when all these great masses, which for the most part had hitherto been in a submarine condition, were elevated and desiccated, so as to constitute *terra firma*, probably in the form of a rocky and abrupt island. It was this land, of whose altitude we can now form no accurate idea, which furnished the pebbles, sand, marl and other materials which compose the molasse and nagelflue. The absence of all links to connect this molasse of the Northern Alps with the pre-existing eocene strata coincides, therefore, with the facts, that, owing to disturbance and elevation, the older tertiary strata constituted terrestrial masses, before the earliest-formed pebbles or sand of the nagelflue were deposited. In this way the vast hiatus between the one set of rocks and the other is well explained. In examining the molasse, we are assured by its fossil remains, whether animal or vegetable, that during the very long period which must have elapsed during its increment, the climate must have been much warmer than that of the same region in the present day. The arborescent palms and intertropical plants which then grew upon the adjacent lands of the Alps and the Jura, the rhinoceros and other large herbivorous quadrupeds which browsed upon them, the large tortoises and the great aquatic salamanders of the lakes, as well as the marine shells of the then bays of the sea, are all unanswerable evidences of a very different climate from that which now prevails. So far we can without difficulty picture to ourselves the former state of things during the accumulation of the molasse. But when we attempt satisfactorily to analyse the physical changes even of this æra, we encounter considerable difficulties. The boldest speculator may be well startled when he is called upon to explain the *modus operandi* by which regularly stratified masses, thousands of feet thick and for the most part formed under fresh water, have been piled up one on the other. He may at first suppose that the well-rounded Alpine pebbles in these strata resulted from the action of various rivers; but a survey of the region soon convinces him that such local causes would be wholly inadequate to explain such a general phænomenon. The grandeur, width, depth, and, above all, the longitudinal persistence of this enormous mound of detrital, yet finely laminated materials, ranging as it does *along the whole external northern face of the chain*, can never be explained by the action of separate rivers which issued from openings into insular lands, unquestionably of much less height than the present Alps. Such lands could only give rise to small partial deltas, each streaming out from the centre of their origin like spokes in a wheel, and could never have produced the one gigantic accumulation of the molasse and nagelflue, which does not run far up into the recesses of the Alps, but constitutes, on the contrary, their broad, external barrier. It may, indeed, be suggested that the detritus resulting from innumerable small torrents descending from a precipitous rocky isle, were accumulated on a steep shelving shore, like that of the present Maritime Alps; but however this may have been, it is manifest that the bottom of the waters which bathed that shore, whether freshwater, brackish or marine, must have been successively depressed to enormous depths. This long-continued depression can indeed alone enable us to account for the heaping-up of these subaqueous materials throughout such a thickness, and consequently during so long a period. It is also self-evident, that whilst they were depositing, the materials of the molasse must have been arranged in strata which sloped away from their parent rocks of the Alps.

At this point, then, in the history of these mountains, we can arrive by an interpretation of the materials in our hands. But with the close of the molasse period, a change came over the surface, compared with which all antecedent phænomena fade away in importance. The very deposits of molasse and pebbles, which till then formed sloping deposits on the shore, or more or less grand horizontal masses at certain distances from it, suddenly underwent those powerful upheavals parallel to the lines of dislocation of the adjacent chain;—movements which not only threw up horizontal strata into vertical axes, but cast down the youngest accumulations of that long period into positions which make them appear to pass under the very rocks out of which they had been formed. Although in estimating such gigantic movements the powers of imagination are at fault, surely it is not unphilosophical, with such unanswerable data before us, to believe that in those days the crust of the earth was affected by forces of infinitely greater intensity than those which now prevail. That the elevation, dislocation and apparent inversion of the molasse was a sudden operation or catastrophe, is clearly demonstrated, both by the physical relations of the strata of that age to those which succeeded them, and by proofs of an immediate change of climate, probably due to a great rise of new lands, and the elevation to much higher altitudes of all the country which preexisted. In attestation of both these inferences, we see the ends of the inclined, and often vertical beds of molasse, whose contents bespeak a warm or Mediterranean climate, covered abruptly by horizontal accumulations of ancient alluvia, the animals and vegetables in which announce a climate little, if at all, different from that of the present day. The extent to which these old water-worn alluvia once filled up the valleys of the Alps, thereby indicating that the chain was then of less altitude than at present—the formation of ancient glaciers—the transport of huge erratic blocks to vast distances, and the great and irregular elevations and deep denudations which the whole area has undergone, are all phænomena pertaining to this most interesting chain, on which, though much has already been said, I hope at a future day to express my opinions.

Part II.

On the Cretaceous and Nummulitic Rocks of the Carpathian Mountains.

In 1843 I examined the northern flank of the Tatra group of the Carpathian mountains with Professor Zeuschner, and although I never published a detailed account of that survey, I gave the general results of it in the work upon Russia and the Ural Mountains. I then believed that all the Carpathian sandstones, as well as the flysch of the Alps, were of cretaceous age; but I now present a section (fig. 31) accompanied by explanations, to show, that whilst many of these sandstones are of secondary age, there are others which, surmounting true nummulitic eocene deposits, are clearly tertiary. This section is therefore now brought forward, both to confirm what has been stated in the preceding pages, and to extend and modify the view which I previously entertained concerning the classification of the formations on the flanks of the Carpathian mountains†.

The lofty axis of the Tatra is occupied by granitic rocks, which on their northern side are flanked, first by talc schists, and next by hard quartzose and altered sandstones, concerning whose age I will not speculate (see 1 & 2 of fig. 31). These rocks are surmounted by great masses of hard subcrystalline limestone (3), often in a state of marble, and with few traces of regular bedding. Near the iron forges of Zagopane, these limestones are visibly stratified, plunging to the north, and they there alternate with a schistose shale (3*) in which the *Terebratula biplicata* occurs in abundance. Again, in the turreted ridges called Muran (or the Wall) the limestones also dip to

† See Russia and the Ural Mountains, vol. i. p. 264.

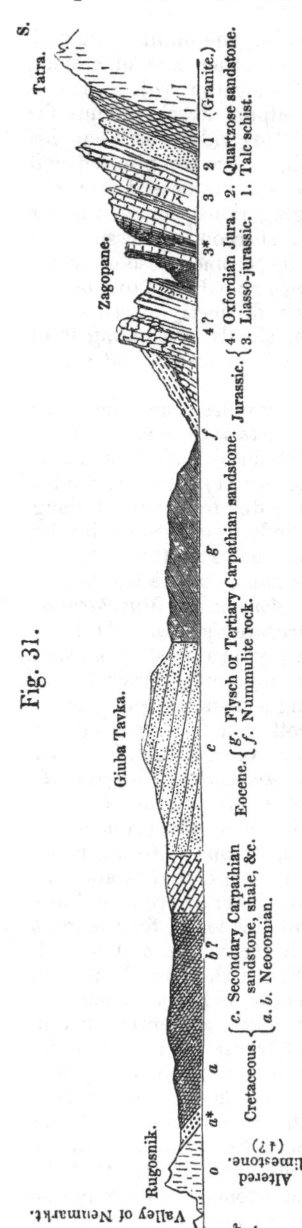

Fig. 31.

the north, but with many undulations and fractures, and in them the following fossils are found, as identified by M. Zeuschner:—*Ammonites Walcotti*, *A. Bucklandi*, *A. annularis*, *Nautilus acutus* (V. Buch), *Belemnites digitalis*, *Terebratula biplicata*, *Spirifer Walcotti*, *S. rostratus*, with Aptychi, Cidarites, Pentacrinites, and some remains of ichthyolites. This group of fossils leaves no doubt, that the limestones containing them belong to the liasso-jurassic limestones. In the interval between the spots where these fossils are collected and the outer edge of the Tatra, there are other limestones in a more or less crystalline state, which, compressed by high inclination into a small horizontal distance, are difficult of access on account of dense woods and their rugged outline. To these I cannot pretend to assign a precise age. On their flank, and particularly on the left bank of the Biala Dujanec, where that stream issues from the gorge of Zagopane, they are unconformably and irregularly covered by a band of nummulitic limestone (*f*), which dips off at an angle of 35° to 40° and passes under a portion of certain schists, sandstones and impure limestones (*g*), which occupy a portion of the hilly tract extending northwards to the valley of Neumarkt. This nummulite limestone is thick-bedded, of grey colour, in part a coarse calcareous grit, and even a small conglomerate made up of fragments of the underlying limestones, and is much charged with magnesia. It contains nummulites throughout a thickness of upwards of 100 feet, but most abundantly in the upper beds. Among these, besides the *Nummulina globulus*, Leym.?, there is the large species *N. planospira*?, so common in the Alps and elsewhere; and these typical fossils are also, as in many other regions cited, associated with certain pectens, ostreæ, &c., and large echinoderms, &c. In short, the fossil assemblage of genera and forms is so precisely the same as that seen in the supracretaceous nummulitic rocks of the Alps, that no doubt can exist as to the age of the deposit. Though denuded at this point, the nummulitic rock is conformably followed to the north, on the other side of a small brook, by dark shale and grey and green sandstone (*g*), which as certainly represent a portion of the upper alpine flysch. Thus far all is clear. But in traversing the undulating ridges between this spot and Wieliczka or Cracow on the north, a very complicated and broken series of sandstones, shale and limestones is passed over, the greater part of which have hitherto been gregariously merged under the name of Carpathian sandstone. Now, as secondary fossils have been found in some of them near Cracow, it becomes absolutely necessary to endeavour to explain the apparent anomaly and to separate the above-mentioned tertiary flysch, which distinctly overlies the nummulitic eocene, from other rocks, often closely resembling it in mineral characters, which are as certainly of cretaceous and secondary age.

Scarcely has the traveller advanced a few miles from the outer edge of the Tatra to the north, than he meets with a low ridge of limestone, which runs parallel to the main chain by Zafflary and Rugosnik. On inspecting this limestone (*o* of fig. 31) I had no doubt, that its mural form and altered condition were due to an upcast along a line of eruption. This supposition was, indeed, confirmed by the existence of an outburst of porphyry a few miles to the east, precisely on the strike of the beds. From the names of the fossils first collected by Professor Zeuschner, such as *Ammonites Murchisonæ, A. Conybeari, A. biplex, A. Tatricus, Terebratula dyphia,* and others, it appeared probable that this rock was simply an upcast on a small scale of some upper portion of the great adjacent jurassic chain, which had been upheaved through overlying schists and sandstones. More careful examination of other fossils collected from this locality by Zeuschner has, as Count Keyserling informs me, detected at least eight species of the lower neocomian, viz. *Ammonites Calypso, A. Morellianus, A. diphyllus, A. picturatus, A. subfimbriatus, A. fascicularis,* with *Scaphites Ivanii,* &c. In this collocation (even if the names in the first list be correct) there need, it appears to me, be no contradiction; for it is the usual case in the Alps, that strata with Oxfordian fossils are at once overlaid by the lower neocomian limestone. The physical features, indeed, favour this view; for the mass of the lowest limestone visible is a highly altered, veined and reddish rock, in many parts amorphous and crystalline, with many slickenside surfaces, and in parts a breccia, which presenting bluff escarpments to the valley of Neumarkt, is overlaid to the south, as represented in the diagram, by dark shale and nodules of ironstone, and then by thinner-bedded, scaly, greyish-white limestones (*a**), which may well represent the lower neocomian, and in which I doubt not the last-mentioned species were found. I persist, therefore, in my belief that these limestones were really upheaved, along a fissure parallel to the main Carpathian chain†. It is indeed manifest (see fig. 31) that the north and south edges of this trough of sandstones are entirely dissimilar; for the strata constituting its north end rest upon limestones

† See Russia in Europe and the Ural Mountains, vol. i. p. 264.

containing cretaceous and jurassic fossils; whilst its southern limb is composed of nummulitic rocks of eocene age resting on lias.

Although I had not sufficient time at my disposal to determine the detailed relations of all the strata between the external face of the Tatra and the valley of Neumarkt, a section with which M. Zeuschner favoured me sufficiently explains, that the flysch (*g*) which overlies the nummulites, and dips to the north, is met by a great mass of Carpathian sandstone, &c., which occupying the Giuba Tavka, is inclined to the south in the manner represented in fig. 31. Subjected to other fractures, this sandstone (probably its lower member) is found a little further northwards to overlie conformably the calcareous ridge (*o*) before spoken of, in the upper part of which neocomian or lower greensand fossils occur, and whose lower division is characterized by Oxfordian jurassic forms. In this way I can readily understand how the nummulitic rock (*f*) and its overlying member (*g*) should really be eocene, whilst the great mass of Carpathian sandstone (*c*) which is separated from the former by a fault, may, as I have always thought, represent parts of the cretaceous system (upper greensand, &c.?). I can also now well understand how the equivalents of the lower greensand (*a*, *b*) should have afforded the above-mentioned characteristic fossils; the whole reposing on an upcast outer ridge of jurassic limestone. Having convinced myself that the nummulitic rocks on the north flank of the Tatra are eocene, I cannot doubt that the masses thereof which lie on the south side of that chain are of similar age. Thus, the sections of M. Zeuschner would lead me to believe, that the nummulitic rocks and overlying sandstones of the valley of Kradak, which he considers to be jurassic, are really the true eocene, which there reposing on lower Jura are truncated against the granite of the higher Tatra. Other nummulitic rocks are repeated to the south of the lesser Tatra.

For all these reasons I feel assured, that Professor Zeuschner has erred in placing this nummulitic limestone in the jurassic series, or at the base of the whole external zone of the Carpathian sandstones and limestones. For, whatever may be the age of the formation on which this nummulitic rock reposes, its zoological characters are unmistakeable; whilst both on the north and south sides of the Tatra, it is immediately covered by strata which represent the flysch.

In the Carpathians, as in the Alps and Italy, great confusion has arisen from deciding on the age of sandstones by their mere mineral characters; for, although it is manifest that rocks of this aspect and containing fucoids clearly overlie the nummulitic limestone, there are other cases, like those of Gosau and other parts of the Eastern Alps, where the lithological characters of the eocene greensand descend far into the cretaceous system. The highly contorted, broken and dislocated region between Neumarkt and Rugosnik on the south and Cracow on the north, over which I passed, affords a good illustration of this point, and also of the extreme difficulty, in the absence of fossils, of being able to draw any neat line of demarcation between some of these sandstones and schists. I have repeatedly noted, that the presence of fucoids can never be accepted as any test of the age of rocks; inasmuch as these fossils have a vertical range from the secondary greensand upwards to beds thousands of feet above the equivalent of the white chalk. In the Carpathians, the same or nearly the same lithological character prevails from the strata representing the lower neocomian up into beds above the nummulitic limestone; and if the normal relations of this region were not excessively distorted, we should, I have no doubt, see that the younger secondary and older tertiary there often pass into each other, preserving the same mineral types.

Thus, if the section (fig. 31) from the flanks of the Tatra be continued from the environs of Zafflary and the valley of Neumarkt to the valley of the Vistula, we may interpret the sandstones, conglomerates and schists of that tract to be both of secondary and tertiary age; for, notwithstanding many dislocations and contortions, it appeared to me, that on the whole the grey sandstones of the hills north of the Biala Dunajec dip away from the axis of Zafflary and Rugosnik, in which jurassic and lower neocomian fossils occur. In this way the hills near Svienty Cruz, about 2000 feet above the sea, which contain fragments of coaly matter and thin seams of lignite, may possibly be paralleled with the strata, which at the outer zone of all the series of similar aspect, extending from Liebertoft to the hills south of Wieliczka, contain what are called upper neocomian, or lower greensand fossils. However this may be, one of the underlying masses dips to the north, and the other or outer zone to the south. The result is, that the greater portion of the intervening strata lie in a rudely undulating and broken trough; and thus I am disposed to think, until contradictory fossil proofs be obtained, that a portion of the series north of Svienty Sebastian, consisting of thick-bedded macigno sandstone of grey and grass-green colours with white veins, (which at Struya and in the hill of Kotan near Luboin are surmounted by dark shale and schist,) and also the strata extending to the valley of the Rabba, may stand in the place of the supracretaceous macigno alpin or upper flysch of the Alps.

But wherever the white chalk or its representative and the nummulitic limestone are absent on the flanks of the Carpathians, and fossils cannot be detected, the geologist must be at fault. Fortunately, however, the palæontological researches of Prof. Zeuschner have proved, that many of these green sandstones, schists and conglomerates are of true cretaceous age. Thus, in the ravines near Liebertoft, two Polish miles south of Cracow, M. Zeuschner led me into gorges where grey, flaglike calcareous grits, occasionally passing almost into conglomerates with black fragments (lithologically, indeed, undistinguishable from supracretaceous rocks), contain ammonites, belemnites, and a species of Aptychus. In these strata, as they range thence towards Wieliczka, M. Zeuschner has since detected *Belemnites bipartitus* (Blainv.), *B. pistilliformis* (Bl.), *B. dilatatus* (Bl.), which present on the whole adequate proof that these sandy beds represent the neocomian. In short, they resemble to some extent the English type of lower greensand. In the county of Trentschin, between Orlova and Podkrad, the so-called Carpathian sandstone, on the other hand, is probably the upper greensand; for, along the space of half a German mile, it there contains *Gryphæa columba, Cardium Hillanum*, and in other places (Zips, Zglo, Wercizer) *Pholadomya Esmarckii*; whilst plants referred by Göppert to the upper greensand, including *Salicites crassifolius*, occur at Kluknawa and Petzoldtii.

It is thus manifest, that in the generic word Carpathian sandstone, as in the words "Wiener sandstein," "flysch" and "macigno," deposits of lower greensand, upper greensand, and of supracretaceous or eocene greensand have been confounded. The value, therefore, of the section between Zagopane and Zafflary (fig. 31) is apparent, because the order of superposition there clearly establishes a parallel between the schists and sandstone overlying the nummulite rocks and the great mass of strata of that age in the Alps. On the other hand, the fossil researches of M. Zeuschner afford clear evidence, that other and large portions of this sandy argillaceous series are equivalents of members of the cretaceous system*. This is precisely what I have indicated, where the mineral representatives of the white chalk of Switzerland and Bavaria approach the Eastern Alps, and where the whole series between the neocomian limestone and the molasse or nagelflue assumes its arenaceous or northern type.

Part III.

On the chief Formations of the Apennines and Italy.

Although less complicated than the Alps, and not containing a vestige of the older formations which have been detected in parts of that chain, the Apennines and their flanks offer many difficult problems, which even at this day remain in some degree obscure. The labours of native geologists in the last few years have, it is true, done much to clear away these doubts, the proofs of which will be found in the proceedings of the last three meetings of the men of science of Italy†. After personal inspection of some of the tracts in which the leading questions have been agitated, I will now endeavour to point out the extent to which the structure of the peninsula agrees with that of the Alps. In the first place, then, the whole of the palæozoic series is wanting in the continent of Italy, nor are there sufficient grounds for supposing that the trias has any existence in it. For although the Marquess Pareto, one of the leaders of our science in Italy, has suggested, that the conglomerate called Verrucano (the

* Count Keyserling, to whom I am indebted for a clear and concise view of the last researches of Prof. Zeuschner, reminds me, that the curious body called a Sphærosiderites, and described by Professor Glocker of Breslau (Acta Acad. Cæs. Leop. Carol. Nat. Curios. tom. xix. part 2. p 673 and tab. 79), which M. von Buch first proclaimed to be a Nautilus, has under the examination of M. von Hauer, jun., proved to be the *Nautilus plicatus* (Fitton) of the lower greensand, or *N. requienianus* (D'Orb.). This fossil occurs in Moravia near Frankstatt and Tickau, in what is called by M. von Hauer "*Wiener sandstein*!" For a full account of the secondary fossils of the Carpathians, see Zeuschner's memoir on the structure of the Tatra Verh. R.K. Miner. Gesr. St. Petersburgh, 1847, which I only received when this sheet was in print. So far as I have had time to study it, this memoir would rather confirm my opinions.—June 25.

† See the volumes of the "Riunione degli Scienziati Italiani" of the Milan, Genoa and Venice meetings.

lowest known sedimentary stratum) may be the equivalent of the trias, that opinion cannot be adopted without some fossil evidences. Whilst the mainland is void of palæozoic rocks, they exist however in Sardinia, where they were discovered by General della Marmora. These rocks prove by their fossils to be true Silurian limestones and schists *. In Sardinia, deposits with species of palæozoic coal plants also occur, but unluckily the political troubles of Italy prevented my examination of these rocks. General della Marmora has, however, left no spot of the island unsurveyed, and having made a beautiful topographical map of it, he will soon complete his important work, and inform us whether the coal deposits of that island, like those of Oporto described by Mr. Sharpe, are associated with the Silurian rocks, or are of subsequent age. The existence of Lower Silurian rocks in Portugal, as recently made known to us by Mr. Sharpe †—the prevalence of Devonian fossils in the north of Spain, and the presence of both Silurian and Devonian strata in Morocco ‡, where they were first recognised by M. Coquand—their persistence along the African Atlas and their reappearance near Constantinople, are data sufficient to enable us to picture to ourselves a vast girdle of palæozoic rocks of which the Alps and the Pyrenees formed the northern and the north-western limits, and which, having been elevated from beneath the sea at a very ancient period, have constituted the shores of a large Mediterranean in which the secondary and tertiary rocks of Italy have since been accumulated. In this sense, Sardinia may perhaps be only viewed as a detached island in this ancient basin.

Excluding from our present consideration the eruptive rocks, whether plutonic or volcanic, which have perforated the subsoil of Italy in so many places, and not now alluding to certain ancient crystalline rocks of Calabria and Sicily, it may be said that the chief mineral masses of the peninsula in ascending order are—1st, limestones and schists; 2ndly, hard sandstones and impure limestones often compact; and 3rdly, marls, sands and conglomerates. The lowest of these great lithological groups embraces in some regions both the jurassic and cretaceous systems; the second or arenaceous group represents in given countries both the upper cretaceous and that which I have shown to be the eocene of the Alps; whilst the third contains the miocene, pliocene, and other overlying strata. To this general litho-

* General della Marmora kindly sent to me a collection of those fossils, including orthoceratites and graptolites. These were examined and partially named by M. de Verneuil, to whom I transmitted them; but having lost that memorandum, which was forwarded to me in Italy, I subsequently referred these fossils to Mr. Sharpe, who is satisfied that they belong to the Lower Silurian group. In addition to orthoceratites, graptolites, crinoids, &c., Mr. Sharpe recognises eight or nine species of Orthis, among the best-preserved of which are the *Orthis patera* (Salter, MSS.), common near Bala, and *O. Lusitanica* (Sharpe), of the Lower Silurian rocks of Vallongo near Oporto, closely related to the *O. flabellulum*, Sow. Sil. Syst. The *Spirifer terebratuliformis* (M'Coy) also occurs,—a species of the Lower Silurian rocks of Cumberland and Ireland.

† See Journ. Geol. Soc. Lond. vol. v. p. 142, and pl. 6. fig. 5 (Russia in Europe and Ural Mountains, Map of).

‡ For the Devonian rocks of the Asturias see Paillette's memoir, Bull. Soc. Géol. de France, vol. ii. p. 439, and for the palæozoic rocks of Morocco see id. vol. iv. p. 1188.

logical arrangement there are, however, local exceptions of considerable extent. In many tracts, also, the absence or extreme rarity of fossils, and the rapid undulations and frequent breaks in the strata, render it very difficult closely to identify each rock-formation with its equivalent in other parts of Europe. This last remark applies chiefly to the rocks of jurassic and cretaceous age; for the great masses of the hard sandstone or macigno, particularly where it is associated with nummulites, are unquestionably supracretaceous or eocene; whilst each of the younger deposits which overlie them is easily referable, both by order of superposition and abundant organic remains, to its respective place in the tertiary series.

In his recently published general geological map of Italy and its adjacent isles, M. Collegno having inserted the Silurian rocks of Sardinia as the lowest known sedimentary deposit, attempts two divisions only of all the rock masses of the peninsula beneath the miocene. The lowest of these, from the verrucano conglomerate upwards to the Oxfordian Jura, is coloured as jurassic; whilst all the overlying strata, whether they represent the neocomian, upper greensand and chalk, together with the nummulitic limestones and macigno, are classed together as cretaceous. It will be readily understood, from what has been said of the Alps, that I must object to the arbitrary union of the two last-mentioned masses with the cretaceous rocks; and hence one of my chief objects will be to show, that in Italy, as in the Alps, the nummulitic and upper macigno group is also of eocene age. I shall further indicate the existence of a natural succession from the top of the eocene or bottom of the miocene up into the pliocene, wherein the fossils exhibit a zoological transition into the latter period. But before I enter on these prominent points of this part of the memoir, I will first say a few words on the jurassic rocks, which are the oldest in which organic remains have been discovered in the peninsula, and then give a brief sketch of the true cretaceous rocks which succeed to them.

The best key to an acquaintance with the lowest strata containing organic remains, with which I am acquainted, is that which is exhibited in the promontories of the gulf of La Spezia and the adjacent parallel ridges of the Apuan Alps. This tract has long been known to English and French geologists by the able description of Sir Henry De la Beche, published many years ago, and to Italians by a memoir of M. Guidoni *. Even at the time when he wrote, Sir H. De la Beche suggested, that the fossils found on the west side of the bay were probably of oolitic or jurassic age, though from their peculiar characters and the supposed presence among them of orthoceratites and goniatites (the latter being then called ammonites of the coal-fields), he very properly left the question somewhat open. But since then, additional collections and a rigid examination of the organic remains have settled the question. The supposed small orthoceratites are found to be simply the alveoli of belemnites, and the ammonites, though not occurring in England, belong to forms known in the jurassic series of Southern Europe. Grouping the observations of his contemporaries, and adding fresh data of his own, the

* See Trans. Géol. Soc. Fr. 1ère sér. vol. i. p. 23, and Giornale Ligustico, 1828.

late Professor Pilla* has recently given a section across the gulf of La Spezia, with the main points of which I entirely agree, particularly in proving, that the ammonitiferous band is not the oldest limestone of the tract, as was formerly supposed, but, on the contrary, is the youngest of its Jura deposits. In any attempt to classify the jurassic rocks of Italy, it must be admitted, that they differ so much from the types of Northern Europe, whether in the composition of the rocks or in the paucity and peculiarity of their fossils, that English geologists will agree with me in thinking, that it is even more unwise than in the Alps to endeavour to force their divisions into too close an accordance with our well-known and clearly characterized British formations. The true doctrine on this point has, indeed, been clearly laid down by Von Buch in a masterly generalization, in which, dwelling on certain marked fossils only, which pervade the Alps, Carpathians and Italy, he has signalized the existence of two great bands, the lower of which may be termed Jura-liassic; the other and overlying mass, the equivalent of the Oxford oolite and clay. I have already indicated in a general manner how this classification is applicable to the Alps, and I have now only to add, that though it has been as yet much less clearly developed in Italy, there are sufficient evidences of its value among the undulations of the Apennines and their flanking parallel ridges.

Jurassic formations in the gulf of La Spezia, in the adjoining mountains of the Apuan Alps, and in the Monti Pisani, &c.—The promontories which flank the long and deep bay of La Spezia on the E. and W. are composed of limestones, which, trending from N.N.W. to S.S.E., are parallel to the loftier ridges of the same age, which further in the interior and to the south constitute the serrated chain of Carrara, Massa and Serravezza, and are, after a short interval, reproduced in the Pisan hills. After looking at the latter I walked over the Apuan Alps, passing from Gallicano in the valley of the Serchio on the east, by the peaks of Le Pannie and the pass of Petrosciano to Stazzemma and Serravezza; and then flanking the western zone of these Alps by Massa and Carrara to Sarzana, I traversed to La Spezia. If I had seen the calcareous masses in the Apuan Alps only as they there appear in the form of dolomites, rauch-kalk, and many varieties of ornamental and statuary marble, I should have been wholly unprepared to admit that they could be the equivalents of the liassic and jurassic series. But I satisfied myself that all these crystalline rocks, even where they rise into the lofty peaks of Altissimo, are simply altered masses, which in their prolongation to the Pisan hills contain fossils. Among the lowest strata are crystalline schists and pebbly conglomerate or verrucano. The geological equivalent of this conglomerate has been much discussed; some geologists, like M. Pilla, desiring to prove it to be palæozoic; others, like Pareto as before said, believing it to represent the trias; and others, including Collegno, viewing it simply as the base of the lias. In the mean-

* Unhappily cut off in the flower of his age at Mantua, in the late war in the north of Italy. I had not the advantage of being acquainted with the memoir of Professor Pilla when I examined La Spezia, but I was aware of his views in general, as also of those of M. Collegno, who communicated a description of the tract to the meeting of the talian men of science at Venice, in which he specially adverted to a great longitudinal fault.

time, no trace of organisms having been detected in the verrucano, it is enough to repeat, that it underlies the *fossiliferous* liasso-jurassic limestones. The lowest masses in the deep gorges near Stazzemma are the well-known mottled "Bardiglio" and other marbles. These are overlaid by schists with quartz veins, which have been converted into dark slates having a true cleavage, and are largely worked for use. The latter are covered by massive buttresses of cavernous "rauch-kalk," in parts graduating into a black and dark dolomite forming picturesque peaks. On the eastern side of the range this massive buttress is irregularly overlapped by lighter-coloured limestones with flints, possibly the representative of the neocomian limestones, which in their turn throw off macigno and other overlying rocks. In ascertaining that the crystalline marbles of Carrara are really altered jurassic rocks, Professor Pilla has shown, that the dark-coloured fossiliferous limestone of the valley of the Tecchia, which contains the same fossils as the marble of Porto Venere, can be followed until it graduates by a change of colour and crystallization into the pure white marble of Carrara*. Professor Savi admits that the mineral masses exhibit the same general succession in the Pietre Santine and Apuan Alps as in the Pisan hills†. Now the latter, which I examined, are unquestionably for the most part of jurassic, or of what some geologists may call Jura-liassic age, as proved by fossils.

In the parallel of Carrara and to the north of that place, the white marble rocks, forming regular strata, rise up with associated schists into the lofty peaks of Altissimo, &c., and dipping to the W. and N.N.W. form the eastern boundary of a great trough watered by the Magra, the centre of which, occupied by tertiary and alluvial accumulations (Caniparolo and Sarzana), is flanked on either side by low hills of macigno, the strata of which repose, on the east, upon the limestones of the Apuan Alps, and on the west, upon the calcareous promontory that forms the eastern side of the gulf of La Spezia.

When this promontory is surveyed in the coast cliffs between the headland of Ponte Corvo near the mouth of the Magra and the town of La Spezia, it is seen to be made up, on a miniature scale, of nearly all the varieties of limestone, schist, breccia, rauch-kalk and marble, which constitute the lofty parallel chain of the Apuan Alps. I made a detailed examination of all the strata from the south of Capo Corvo, by Porto Telaro to the old fort of St. Bartolommeo, and found that there was there the same ascending order as in the Apuan Alps, and I was therefore convinced that it was simply a less elevated parallel fold of similar rock-masses (fig. 32). The lower strata are grey, calcareous schists, courses of scaly limestone, in parts highly altered, overlapped by a strong band of white, thick-bedded, impure, statuary marble, with a schistose lamination. This passes up into a concretionary, mottled, purple and white limestone, large calcareous geodes being arranged in the laminæ of deposit in a base of glossy purple and green schist. This calcareous group (1) is

* Bull. Soc. Géol. de France, 1847, vol. iv. p. 1069.

† Considerazioni sulla struttura geologica delle Montagne Pietre Santine, dal Prof. Cav. Paolo Savi: Pisa, 1847.

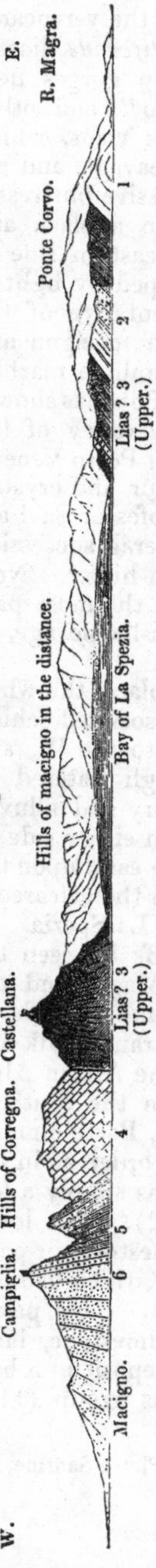

Fig. 32.

W. Macigno. Campiglia. 6 Hills of Corregna. 5 4 Castellana. Lias? 3 (Upper.) Hills of macigno in the distance. Bay of La Spezia. Lias? 3 (Upper.) 2 Ponte Corvo. 1 R. Magra. E.

surmounted by a quartzo-schistose mass (2), which presents the aspect of having undergone a metamorphosis which has affected the hard purple schists, the conglomerate or pebbly beds, and the green schists above them, the latter being traversed in many directions by white veins of carbonate of lime. This mass (2), not less than 300 feet thick, is amorphous, and in colour partially resembles serpentine*. After passing a portion of the cliff which is obscured by detritus, a dark or almost black limestone with white veins appears, undulating irregularly, and plunging on the whole to the west and by north. This rock is covered by dark schists, the whole probably representing the dark limestone and schists of Porto Venere on the opposite side of the bay. These thin-bedded dark masses are followed by the remarkable rocks which constitute the sea-worn, amorphous and cavernous rauch-kalk on which the picturesque old town of Porto Telaro is built. From this point, passing to the fort of St. Bartolommeo, there are undulations and breaks in lower and obscurer cliffs, in which the thin-bedded siliceous schists (slates of Stazzemma) appear here and there beneath the rauch-kalk. All the calcareous rocks of this series are flanked by a vertically twisted and confused, coarse conglomerate, made up of lumps of all the above-mentioned rocks.

I have spoken of this highly modified range on the east side of the gulf to show its lithological accordance with the chief masses in the Apuan Alps, and because it exhibits the same order of succession of mineral masses. It is however only on the western shores of the bay, in another parallel undulation of these limestones, further removed from the chief axes of disturbance, that their age can be read off by help of some imbedded fossils. The black limestones, with white and yellow veins and associated dark schists (No. 3), but not so metamorphosed as on the east side of the bay, form the chief nucleus of this western promontory. Ranging in highly inclined and vertical forms, by the lofty, unfinished fort of Castellana, they strike from N.N.W. to S.S.E., into the isle of Palmaria, where they are largely quarried as the black and brown marble of Porto Venere. Among the fossils in this rock I procured a *Lima*, which resembles a lower secondary fossil, and certain imperfect

* I could not help suspecting the contiguity of some eruptive rock to this peculiar limestone, and my boatman assured me that when the sea was lower one of my predecessors had discovered a point of porphyry.

coralline bodies, which occasionally weather out on the surface. M. Pilla compares this rock with the lower ammonitic or liasso-jurassic limestones of Como, and he cites the fossils as pertaining to the genera *Cardita, Modiola, Pecten, Terebratula,* but does not name the species.

A transverse section of the western promontory, from the gulf south of La Spezia by the hills of Corregna to Campiglia, exhibits a great line of fracture*, irregularly parallel to the ridge, along which the highly twisted beds of dark-coloured limestones and schists (4) have been snapped asunder, and by which a portion of the Porto Venere series is thrown into an inverted position, and seems irregularly to overlap a series of strata which are unquestionably of younger age. These are, first, grey limestones and dolomites, dipping to the N. and S. of E. at 45°, followed by schists, shale, and very thin-bedded, finely laminated *red* and grey limestone, the angle of inclination in which increases gradually to 70° E.N.E. and E. It is in this group, particularly in certain red and grey limestones, that most of the ammonites and other peculiar fossils of La Spezia occur, which have been enumerated by Sir H. De la Beche.

The above fossil zone is underlaid by rotten schists with sandy rotten limestone, and then by numerous alternations of green and grey limestone, whitish calcareous grits, purple and white and red schists, in parts almost jaspideous, with courses of whitish limestone, the whole (5) in very thin-bedded strata, in which unluckily no fossils have been discovered. This calcareous series is flanked by a wall of sandy and pebbly conglomerate, on which stands the lofty village of Campiglia; and the beds, after first positively underlying all the older series at an angle of 80°, first become vertical, and then dipping away to the west form the base of all the hills of fine macigno sandstone. This conglomerate and the associated macigno are thus seen to partake intimately of the same great elevations and flexures which have affected the older limestones in the Apuan Alps and in the gulf of La Spezia.

It is now well known that the macigno of this tract, which, both from mineral character and order of superposition, was formerly taken by geologists for the most ancient, is in fact the youngest of this series; but whether it represents a portion of the cretaceous system, or is younger, is, in the absence of all fossils, still doubtful. That this macigno is in an inverted position is also noticed by Pilla†. In enumerating the list of fossils of the upper jurassic at Corregna, that author mentions *Ammonites Tatricus* and *Nerinæa* (the former I found myself), and I have therefore no doubt that this band represents the "ammonitico rosso" or Oxfordian of the Alps, and that the masses intercalated between it and the macigno are probably imperfect and non-fossiliferous equivalents of some member of the cretaceous rocks.

* M. Collegno sent a memoir on La Spezia to be read at the meeting of the Scienziati Italiani of Venice, 1847, in which he indicated a great line of fault parallel to the strike. He afterwards explained his view of the phænomenon to me.

† Bull. de la Soc. Géol. de France, vol. iv. p. 1069, section pl. 6. fig. 2.

I will only add, that I believe the deep bay of La Spezia has been excavated in a synclinal trough of macigno, whilst the hard inferior limestones and marbles have resisted, and form broken anticlinals on either side of the bay. On the east we see, in fact, another and broader parallel synclinal trough, the valley of the Magra, which has been excavated in similar macigno, covered by some remnants of overlying tertiary. Again, when we traverse the great anticlinal of the Apuan Alps, and descend into the valley of the Serchio, we meet with a third and similar trough, on the eastern side of which the lower portion of the macigno rises up into mountainous elevations, of which hereafter.

I have commenced with this superficial sketch of the general arrangement of the rocks in this northern tract of Italy, not merely because the oldest known limestones in the chain of the Apennines are brought out in parallel anticlinals, but because the outline of *ridge and furrow*, here so clearly developed, is a key to the general structure of the Apennines. In truth, the Italian peninsula is not characterized by one central backbone or central axis, but is made up of a frequent repetition of axes, the rocks composing which are sometimes much altered, often dismembered, and frequently covered over by younger sediments; the older portions of the series being only seen at intervals as we follow the chain from N.N.W. to S.S.E. Thus, jurassic rocks have not been detected in the broad and mountainous undulations which extend over the principalities of Lucca and Parma, or the country of Genoa, a region almost entirely occupied by limestones and macigno sandstone, whose age I shall presently endeavour to explain.

In the northern portion of the Tuscan Maremma, I examined, in company with M. Pilla and M. Coquand, the axis of jurassic rocks which, at Campiglia, are converted into domes of crystalline white marble in the contiguity of points of granite. This marble throws off on its flanks a compact thin-bedded limestone with encrinites overlaid by schists, with *Posidonia Jossiæ*, the latter being abruptly truncated against masses of macigno and alberese. This jurassic group is evidently the prolongation of one of the zones of the above-described Apuan Alps and Pisan hills*. The existence of true jurassic rocks has further been clearly indicated by M. E. de Vecchi in Monte Cetona, on another parallel between the Maremma and the Apennines. The nucleus of this hill is evidently the same Oxfordian limestone with *Ammonites Tatricus* which constitutes the upper group at La Spezia. Again, in another parallel, the group of mountains between the Maremma and Sienna, composed of white, yellow and red marbles (the Montagnuolo Senese), the whole reposing on a conglomerate, probably represent the jurassic series, since they are overlaid by scaglia; but no fossils have yet been detected in them.

In following the chief ridges of the Apennines to the south, buttresses of true jurassic rocks are indeed here and there visible, rising out from beneath overlying formations. Ammonites, of the group

* Some of the phænomena on the flanks of the granite of Campiglia and the promontory of Piombino are analogous to those of the adjacent isle of Elba, and the rocks are loaded with various crystals of iron ore.

of *A. elegans*, occur in the mountains east of Perugia, and also in the red marble limestone of Monte Malbe, west of that city. The *Ammonites Tatricus* and others of the Oxfordian group are found in the limestones of Monte Subasio, east of Assisi, and at La Rossa, between Fossato and Fimbriano. In this region also, as in the Apuan Alps and at La Spezia, such ammonitic rocks rise out as axes, throwing off troughs of younger strata, the sides of the hills being for the most part occupied by vast accumulations of macigno.

At Cesi, near Terni, on the very outer western edge of the Apennine chain, I found red limestone and shale, in parts undistinguishable from the "ammonitico rosso" of the Venetian Alps or that of La Spezia, presenting a bluff escarpment to the tertiary subapennine accumulations of the valley of the Tiber, and loaded with characteristic ammonites, such as *Ammonites Tatricus*, *A. biplex*, &c. The red ammonitic rocks of Cesi, which are clearly of Oxfordian age, repose upon a grey limestone of perhaps a thousand feet in thickness, with siliceous or flint nodules, and are covered by flaglike limestones and bosses and peaks of dolomite. Now, if characteristic fossils were not found in the red zone (a rare phænomenon in the Apennines), who could have divined the age of these rocks? and how should we have seen speculations on the underlying flinty cream-coloured limestone being perchance neocomian or scaglia? It is indeed a nauseous task for the geologist to wander for days in these mountains without the trace of fossils, and hence the ammonites of Cesi are invaluable landmarks.

The palæontological researches of Professor Ponzi in the eastern half of the Papal States, when combined with the mineralogical descriptions of his distinguished coadjutor Count Lavinia Medici Spada, in reference to the volcanic and crystalline rocks, will, I trust, at no distant day be embodied in a work which, coupled with the labours of Count A. C. Spada and Prof. Orsini on the Adriatic side of the great axis of the Apennines, will throw much light on this subject. Professor Ponzi has assured me, that however difficult to separate these limestones lithologically, there are numerous places along the western edge of the chief ridge of the Apennines (extending from Scheggia and Monte Cucco on the N.N.W. by Fossato Gualdo to Col Fiorito on the S.S.E.), where limestones with jurassic ammonites occur, and that near La Scheggia and elsewhere these are seen to pass under cretaceous rocks. Now, this ridge is parallel to others of similar age: 1st, that of Monte Subasio and its prolongations on the east side of the valley of Spoleto; 2ndly, of Cesi, west of Terni; and 3rdly, that of Monte Cetona; and another might even be enumerated in the Tuscan Maremma. These facts sufficiently indicate the prevalent outline of ridges and troughs directed from N.N.W. to S.S.E., into which so large a portion of the peninsula is divided. Patient examination, however, can alone detect the extent to which true jurassic rocks, as defined by the fossils, are separable from those of neocomian and cretaceous age.

Amidst the varieties of marble which abound in the Roman States, there is little doubt that the common red "cottannello," of which the great columns in the facade of St. Peter are made, as well as the "breccia di Simone," are the Roman representatives of the "ammonitico rosso," or Oxfordian.

Of the extent to which jurassic rocks may crop out on various parallels of the grand mountainous undulating region of the central Apennines, which runs down from the lofty Sibilla and Leonessa into the kingdom of Naples, no one is yet perfectly informed; but the researches of Orsini and Alessandro Spada teach us that jurassic rocks reoccur on the eastern side of the axis, the chief elevated points of which are either cretaceous or nummulitic. I satisfied myself, indeed, that the mass of mountains coloured as jurassic by Collegno, which extends from Civita Castellana to Gaeta, is cretaceous, and forms part of the rocks of that age around Naples*. The lowest visible strata, however, in the great promontory on the south side of the bay of Naples, and notably the bituminous limestones of Torre Orlando, are classed by Agassiz as jurassic because they contain the *Pycnodus rhombus*, Ag., *Notagogus Pentlandi*, Ag., &c. Again, certain lower strata of the Val Giffoni, east of Nocera, may be referred to the age of the lias, in consequence of the description of their ichthyolites by Sir P. Egerton, viz. the *Semionotus Pentlandi* (Egert.), *S. pustulifer* (Egert.), and *S. minutus* (Egert.) †.

Cretaceous Rocks of Italy.

Clear distinctions between the cretaceous and jurassic rocks of Italy can, for reasons already assigned, be seldom safely effected, except where the one or the other contains fossils, and can thus be compared with the types in the Alps. At Nice on the west, near Milan in the centre, and in the Vicentine on the east, the flanks of the Alps afford us, indeed, good keys, which explain the order of succession. But we no sooner quit the edges of those mountains and advance into Italy, than we lose for a long space nearly all evidences of true cretaceous rocks as proved by their fossils. We then find ourselves in broad, mountainous undulations of sandstone, schist, and impure limestone, some of which have a striking resemblance to the flysch of the Alps. The geological map of Liguria Marittima, by the Marquis Pareto, extending from Nice on the one hand to La Spezia on the other, and the work that accompanies it ‡, expose the difficulties, which even an able geologist intimately acquainted with his country

* The author of this map is aware of the error, and informed me of it before I visited the tract. In the first effort to map a country such errata are inevitable, and our best thanks are due to M. Collegno for his endeavour to produce the first general geological map of his country.

† See Proc. Geol. Soc. Lond. vol. iv. p. 183.

‡ In the 'Cenni geologici sulla Liguria Marittima,' p. 30–47, Pareto considers all the macigno and alberese of Liguria to be cretaceous or secondary, because it contains fucoids. It is nowhere associated with nummulites. But in respect even to the latter, he classes the lowest great band of them at Nice as also cretaceous, because it succeeds in overlying order to a representative of the chalk in which green grains abound. In truth, however, there can be no doubt that all the nummulite rock of Nice is truly eocene and tertiary, and that it reposes on the equivalent of the chalk with inocerami. As to the non-fossiliferous macigno and alberese of the Genovesato, it is hopeless at present to define their age with pre-

experiences, in handling this part of the subject in such a region. At Nice indeed the succession is clearly shown from neocomian limestone, through greensand and chalk, up to overlying nummulitic strata; but in the whole of the tract east and west of Genoa, the slaty, hard, calcareous flagstones (alberese) and the macigno sandstones are grouped together as a higher member of the secondary series. In following the northern edges of the great plain of Piedmont, a wall of crystalline and eruptive rock subtends the alluvial plains of the Po; and though some representatives of cretaceous rocks flank the Alps and rest upon jurassic limestones to the north of Milan and near Como, I shall not now speak of these, because I have not visited them.

Good types of the cretaceous rocks of the north of Italy occur, however, in that outlier of the Alps with which I am acquainted, the Euganean hills. Separated from the chain by the trough of the lower tertiary deposits of the Vicentine, these hills, long known to consist chiefly of eruptive trachytes and scaglia, or the equivalent of the chalk, have, thanks to their vicinity to the abode of scientific men at Padua, been at length well developed. In his elaborate work, full of lithological and mineral description and views concerning the pseudo-volcanic operations of the region, M. Da Rio has also the merit of having enumerated, with the assistance of Professor Catullo and others, a certain number of fossils*. But these were not so described or grouped as to furnish to any extent geological divisions in the secondary rocks, though on the whole the strata containing Ammonites, Belemnites, and certain Echini (*Ananchytes ovatus*) were separated from the strata loaded with nummulites, which, in following Brongniart, M. da Rio considered to be tertiary. The more recent researches of Pasini and Catullo, and particularly those of De Zigno, show that the calcareous masses, formerly known under the terms of grey, white and red scaglia, are divisible into formations by their fossils; the lowest of these representing the Oxfordian or "ammonitico rosso" of the Alps, with *Ammonites Hommairi* (D'Orb.), *A. biplex* (Sow), and *A. Zignoanus* (D'Orb.). The next or neocomian, forming the base of the cretaceous system, is characterized by the *Crioceras Duvalii*, *Belemnites dilatatus* (Blainv.), *Ammonites cryptoceras* (D'Orb.), *A. Astierianus* (D'Orb.), and *A. infundibulum* (D'Orb.). The next overlying stage is considered to be the "Aptien," D'Orb., and contains the *Hippurites neocomiensis*, with Spherulites and *Ammonites Guettardi*. The uppermost band is the scaglia, or true equivalent of the white chalk, with *Inoceramus Lamarckii*, *Ananchytes tuberculatus*, *Holaster*, &c.

In his description of the "Terreno-Cretaceo" of the Venetian Alps, as exhibited in the Monfenera between Fener on the north and Pede-

cision, void as they are of fossils and perforated in all directions by serpentine and eruptive rocks. It may however be supposed, as suggested in the text, that a part of them may be cretaceous, and a part, on the parallel of strata which elsewhere contain nummulites. The same species of fucoids ranging throughout the eocene of the Alps down into the lower chalk of Northern Italy, are no criteria of age.

* Orittologia Euganea del Nobile Niccolo da Rio, Padova, 1836, with a coloured map and a lithographic profile.

robba on the south, M. de Zigno has fully developed the order and component parts of the "cretaceous system" of Northern Italy. Identifying the "biancone" with the "majolica" of Milan, he shows, by its several species of Crioceras and many species of Ammonites, published by D'Orbigny as neocomian from Provence, that these Italian limestones form truly the base of the cretaceous rocks, and are perfectly to be distinguished from the grey and red scaglia above them, and from the Oxfordian Jura or "ammonitico rosso" beneath them. I can only venture to differ from the Italian geologists when they state that a limestone between the neocomian and the upper scaglia contains nummulites. All the small bodies *supposed to be nummulites**, when seen on the surface of such cretaceous limestones, have proved to be other genera of foraminifera (chiefly Orbitolites) when closely examined. The course of sandstone intercalated between the so-called neocomian and the scaglia or chalk is perfectly in accordance with the section of the Grünten and Bavarian Alps (see p. 204).

In the Euganean hills, then, as in the Venetian Alps, the upper member of the chalk is surmounted by the well-known lower tertiary nummulitic rocks of the Vicentine, in which species of nummulites occur, together with Turbinolia and other fossils. This tertiary group runs into the hills south of Vicenza, which constitutes a part of the eocene accumulations before alluded to. When, however, we quit this Euganean outlier or island, and travel over the intervening plains to the centre of the Apennines, we are, as in Liguria, immersed in that very different mineral type to which allusion has already been made. The limestones, some of which may stand in the place of upper members of the cretaceous system, are traversed by serpentines, and scarcely ever contain the trace of animal organic remains. In vain, for example, does the geologist explore the limestones constituting the chief ridge between Bologna and Florence, or the axis between Liguria and Piedmont; for with a few examples of fucoids only, he can find no fossil base-line whatever in descending order, and

* See "Sul Terreno Cretaceo dell' Italia Settentrionale," Padova 1846. When the Men of Science were assembled at Venice in 1847, I in vain endeavoured to detect a true nummulite found in this cretaceous rock. Among the unequivocal neocomian fossil species of D'Orbigny, cited by De Zigno, are, *Ammonites astierianus*, *A. Guettardi*, *A. macilentus*, *A. Juilleti*, *A. inæqualicostatus*, *A. Grasianus*, *A. Morelianus*, *A. subfimbriatus*, *A. recticostatus*, *A. Matheronii*, *A. Terverii*, *A. bidichotomus* (Leym.), *Belemnites dilatatus* (Blainv.), *B. latus* (Bl.), *Crioceras Emerici* (D'Orb.), *C. Duvalii* (Léveillé), and *C. Da Rio* (Zigno). With these and the *Spatangus retusus* are associated two species of Aptychus. One of the latter has been supposed by Von Buch to occur also in the upper Jura. In his memoir before referred to, M. Zeuschner cites two species of jurassic Aptychi and the *Terebratula diphya* as being associated with many forms of neocomian age in the same band of Carpathian rocks, an anomaly which I have endeavoured to explain, pp. 260, 261.

In his description, M. de Zigno further shows, that the conjoint elevation of the cretaceous and overlying tertiaries has extended from the tract where I first described it to the longitude of Feltre. Whilst these pages are passing through the press, I learn from M. de Zigno that he has in great part carried out the work to which I alluded p. 223.

the vast masses of associated macigno sandstone are scarcely more productive. As to the fucoids found in the Florentine Apennines, they have much too great a vertical range to afford any criterion whatever of the true age of the rocks. Forms said to be similar occur in the grey or lower scaglia of the Venetian Alps, and are, as I have shown, still more abundantly developed in the supracretaceous macigno or flysch of the Northern Alps.

Rare as they are, certain fossils have, however, been found; and the existence of the solitary *Hamites Micheli* of Fiesole, of one ammonite discovered by Mr. Pentland, and of another by the Marchese Pareto, are, in the absence of other countervailing fossil proofs, enough to satisfy me, that there is here a zone, which, in a peculiar lithological form, represents the cretaceous system, as on the north flank of the Carpathians. I consider this group to be the equivalent of the upper greensand and chalk, which has assumed very much the same "flysch" or macigno characters as the cretaceous deposits of Gosau. I am very far, however, from agreeing with Professor Savi, that all the macigno is cretaceous. On the contrary, I am convinced that probably *the largest masses* of that rock, and particularly whenever they surmount or are associated with nummulitic strata, are of eocene age. The beds at Perolla, near Massa Marittima, which contain the Gryphæa figured by Pilla, represent in my estimate the uppermost cretaceous, or band of transition into the lower tertiary rocks.

True equivalents of the neocomian are, as before said, rare in the north of Italy. In the environs of the baths of Lucca, the position and lithological aspect of the mountain called "Prato Fiorito" led me to believe that it was of neocomian age. It is composed of compact cream-coloured limestone (*a* and *b*, fig. 33), with numerous nodules of

Fig. 33.

East of the Baths of Lucca.

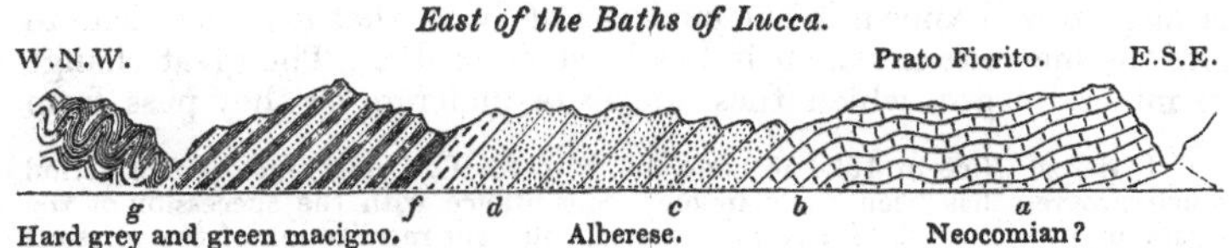

flint, precisely resembling the "biancone" which in the Venetian Alps is proved to be of that age. Moreover, I found it to be surmounted by bands of impure sandy limestone, schist, red scaglia, &c. (*c* and *d*), which might very well pass for the greensand and chalk; whilst the great mass of macigno overlying the whole series constituted the chief peaks of the surrounding mountains. These again are followed by a peculiar calcareous breccia and agglomerate (*f*), which seemed to occupy the usual place of the chief nummulitic zone of Italy; the whole being surmounted by the great mass of macigno of which the surrounding adjacent mountains are for the most part composed. But, after a long search, I could find no organic remains except a rude cast, which might be a Crioceras, and which I detected near the summit of the supposed neocomian. Fucoids indeed are seen on the faces of the scaglia-like and impure limestone which here underlies the great mass of macigno.

Passing from these difficulties, in defining the equivalents of the cretaceous rocks in Liguria, Modena, Lucca and Tuscany, and their relations to inferior and superior strata, the true cretaceous system is not only observable at intervals in the southward range of the Apennines, but on various parallels resumes its calcareous and fossil characters, and constitutes ridges of considerable length in Southern Italy. In the Papal States these limestones, undergoing many undulations and breaks, constitute the chief chains which flank the valleys of Umbria, the Sabine mountains (Tivoli, Subiaco and Palestrina), and the Volscian hills extending to Gaeta and Naples. Although these rocks are in their upper portion chiefly characterized by hippurites, I am unprepared to define to what extent they may be divided into formations representing the neocomian or lower greensand, as separated from the upper greensand and chalk. I will, however, presently describe how these hippuritic limestones of the Sabine hills are surmounted by nummulitic rocks and macigno. In the limestones of Gaeta, whether crystalline, saccharoid or compact, I found many hippurites. The same rocks rising up into the ridge of Monte Marzo, near St. Agata, are underlaid by a thin-bedded, earthy and sometimes bituminous, dark-coloured limestone, which may be considered neocomian. The jurassic limestones of Sorrento are of great thickness and contain hippurites ; whilst the whole of the above series dips under macigno.

On the Eocene Rocks of Italy and their relations.

The group of this age, as clearly indicated by its overlying relations to true cretaceous rocks, has been sufficiently described in respect to the Venetian Alps, the Vicentine and Euganean hills. It is also so well known in the environs of Nice, that it is sufficient to cite the memoirs in which it has been noticed*. The great change in mineral aspect which these deposits undergo, as they pass from

* I visited Nice in 1828 in company with Sir C. Lyell, since which period much progress has been made in our acquaintance with the succession of the strata in its environs. I have, however, a sufficient recollection of the physical relations of the rock-masses to understand the value of the descriptions of the Marquis Pareto (Liguria Marittima), and of M. Perez. The latter gave a very elaborate account of the nummulitic deposits of that tract at the Genoa meeting of the Scienziati Italiani; but whilst he allowed that the greater part of the nummulitic fossils were eocene, still, in compliance with the prevailing fashion, he classed them as cretaceous, as well as the adjacent macigno of the Maritime Alps which overlies the nummulitic group. The sections are, in a general sense, so in accordance with my own in the Alps and Apennines, that it is unnecessary I should do more than say, that they exhibit a succession of various bands containing nummulites, ostreæ, &c., the lowest of which repose on beds (often a greensand) with *inocerami of the chalk*, and which in other places are charged with other types of the upper cretaceous groups. A limestone with neocomian fossils, and another with Oxfordian jurassic fossils, in descending order, completes therefore the analogy with the succession in the Venetian Alps. The conclusion of M. Perez is, that the macigno of the Maritime Alps is everywhere more recent than the nummulitic limestone.

one region to another, and above all the accompanying phænomenon of an almost entire disappearance of organic remains, have necessarily involved them in much obscurity in Liguria, Modena and Tuscany. As a whole, there is indeed a strong lithological resemblance, as before said, between the rocks called macigno by the Italians, and the flysch and Wiener sandstein of the Swiss and Austrian geologists. In the Apennines, as in the Alps (I have already alluded to it in the Apuan ridge), there is a fine-grained small micaceous sandstone which much resembles the ordinary macigno, whose exact age, whether cretaceous or eocene, may be doubtful; but I now simply treat of that macigno which, wherever there have been no inversions, is either intercalated with, or superposed *en masse* to, the nummulitic rocks. If we appeal to the environs of Florence, we see that, however wanting in a clear cretaceous base-line in the vicinity of that city, the whole of the Tuscan series of alberese and macigno sandstones repose upon secondary limestone (chiefly jurassic) in the Pisan hills on the west, at Monte Cetona and Campiglia on the south, and in the central Apennines of Monte Verame and Citta di Castello on the east. In defining the relations of the component parts of this group, I have already expressed my belief, that in parts of Tuscany the lower portion is probably the non-fossiliferous representative of the upper portion of the cretaceous system. In fact, the term "alberese" is so loosely applied to every light-coloured limestone, pure or impure, which dips under or alternates with macigno, and which may happen to contain fucoids, that it would be very hazardous, in tracts so void of organic remains, to define the neat limits between secondary and tertiary. We have not here, as in the Alps, either a neocomian limestone with its fossils to represent the lower greensand of the English, nor anything like the Alpine equivalents of upper greensand and chalk. But, if so obscure in the descending order (and he who crosses the chain from Bologna to Florence will admit this to be the case), these tracts have, however, one strong point of comparison with the Alps in the lithological resemblance of their upper macigno to the flysch of the Swiss. They further resemble certain Alpine tracts in having no rigid boundary or break between the lowest strata in which nummulites occur, and the beds above and below them. The best proof of this is, that Professor Pilla described as one natural physical group, which he termed "Etrurian," that which, on further inquiry, he conceived to be composed in its lower part of strata referable to the chalk, and in its higher part of a peculiar intermediary formation. As this Etrurian system (*so named from the country in all Europe most deprived of organic remains*) is thus composed of both secondary and tertiary strata, it is manifest that the term is geologically inadmissible.

Whilst then the lowest alberese, and some macigno, may remain as very ill-characterized cretaceous rocks, the upper Etrurian of Pilla is in fact nothing more than the eocene group of the Alps, like which it contains, in some localities, zones of nummulitic limestone, and is further surmounted by vast accumulations of macigno sandstone.

The nummulitic limestone of Mosciano, near Florence, having been much spoken of by Pilla and others, and having attracted the notice of the geologists of the meeting at Florence, I visited* and made a section of the strata which I now produce (fig. 34), as it differs

Fig. 34.

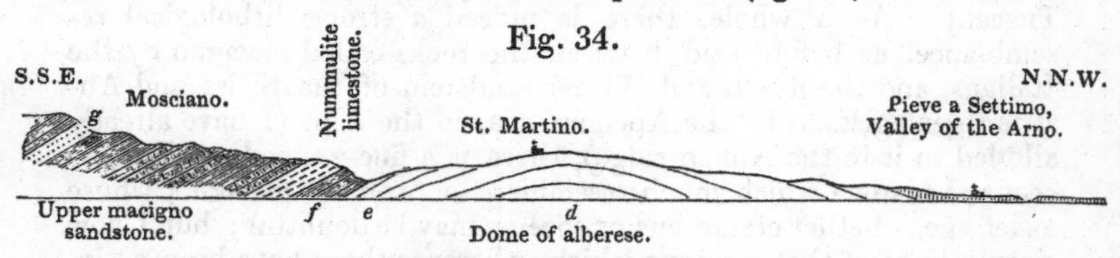

essentially from those hitherto published†. It shows an underlying light grey limestone or alberese, with fucoids followed by schists; next by the nummulitic limestone (*f*); and lastly by a vast overlying mass of macigno sandstone (*g*), as seen in the hills above St. Martino. The lowest beds at the Calcinajo, to the south of Pieve-a-Settimo, are thin-bedded, cream-coloured limestones (*d*?) of conchoidal fracture, with marlstones which contain the *Fucoides intricatus* and *F. Targioni*‡. These limestones, alternating with whitish schists or marl, form a low dome, the south side of which dips under other schists or shales of black and red colours, the "galestro" of the natives (*e*), which are covered by a thin band of micaceous sandstone or macigno. Then come impure gritty light grey limestones in strong beds of four to five feet thick (*f*), which contain small nummulites (*N. globulus*?) and minute foraminifera, and towards their upper portion pebbles of older compact limestone. These graduate upwards into flaglike, sandy, impure limestone of a bluish tint, which forms the passage into a great and distinctly overlying mass of "macigno" (*g*). Beds of coarse grit or small conglomerate occur near the base of the sandstones, in which are pebbles of quartz and diallage, and above them are small micaceous sandstones, which, although weathering yellowish, are of the usual grey macigno colours when freshly quarried. Some of these masses assume spheroidal shapes, and there are other alternations of similar sandstones and shale, and coarse grits (conglomerates, &c.), which occupy the summits of the adjacent hills. Now, all these beds are perfectly conformable, and from below the village of St. Martino to the tops of the hills they dip to the S.S.E. at angles gradually decreasing from 20° to 10°. It is thus seen, that the nummulites occurring in the lower part of all the macigno which is here exposed, are clearly covered by another and much greater mass of macigno. These strata are therefore, according to my view, on the very same parallel as the nummulitic and flysch group of the Alps.

The hills near Pistoja, and, in short, everywhere around the vale

* I was kindly accompanied by the Marchese Carlo Torrigiani and Professor Targioni-Tozzetti.

† Compare my section with that of Pilla in his work, entitled "Distinzione del Terreno Etruri tra' piani secondari del mezzo-giorno di Europa." Pisa, 1846. Tav. iii. fig. 3; and Mém. Soc. Géol. Fr. vol. ii. 2nd ser. p. 163.

‡ Where these limestones are of a bluish-grey colour they are called "Colombino," as distinguished from the whiter beds or "alberese."

of the Arno, afford sections of macigno, and, whether of grey or greenish-grey colours, it is usually the same slightly micaceous and feebly calcareous sandstone with grains of black schist, the fine building-stone of this region. At Ripa-fratta and other places to the north of the Pisan hills, this macigno is seen to dip away from all the chief underlying calcareous masses. But unfortunately the absence of fossils between the ammonitic group of the latter and the lowest beds of true macigno (the interval being occupied by compact limestone with flints), defeats any attempt at close comparisons *.

In ascending the valley of the Arno above Florence, and particularly between Ponte Sieve and Incisa, strong bands of alberese limestone undulate in rapid flexures or anticlinals, and dip under vast thicknesses of macigno, which roll over rapidly to the W.N.W. and E.S.E. At Monte Consuma these macigno rocks contain two or three courses of nummulitic limestone, as Professor Pilla has already indicated.

The grand masses of macigno which occupy the sides of the upper valley of the Tiber near Arezzo and thence range down to the environs of Perugia may be followed to the flanks of the highest Apennines, where they are seen to repose on the secondary limestones. Between Arezzo and Perugia the macigno is copiously developed, forming the hills on the eastern shores of the Thrasymene lake; it there clearly alternates with subordinate calcareous bands, and is itself often slightly calciferous. It is here near the centre of a vast trough, the limits of which are the secondary limestones of Monte Cetona on the west and the Apennines on the east. I was not able to satisfy myself, by any absolute superposition to strata with cretaceous fossils, whether these rocks are really of lower tertiary age; but my impression is that they are simply prolongations of the eocene macigno of Arezzo and Monte Consuma, in which nummulitic bands occur. In examining them I was reminded of an observation made by my lamented friend M. Alex. Brongniart, who, when I first showed him characteristic specimens of the upper silurian rock of Ludlow, exclaimed that they were true "macigno." I assert that the small micaceous, slightly calcareous, earthy sandstones, breaking to a bluish heart within, and weathering to a dirty

* "Sulla costituzione geologica dei monti Pisani, memoria del Prof. Cav. Paolo Savi, Pisa, 1846." Placing the nummulitic limestone as the uppermost bed of the cretaceous rocks, Professor Savi shows that it reposes on alberese with fucoids with and without flints, macigno sandstone, argillaceous schists, and mottled limestone with fucoids. Beneath this upper group are other and darker-coloured limestones, with flints and fucoids, which form the base of the cretaceous rocks. He then classes in the upper member of the jurassic series a light grey limestone (also with flints), which, he says, passes into and contains some of the fossils of the red ammonitic rock, about which no doubt exists. His lower jurassic or lias is made up of whitish limestone with fossil bivalves and turriculated shells, of a dark grey limestone, also with some obscure fossils, and lastly, at its base, of the "Verrucano." The only horizon clearly marked by its secondary fossils in all this series is the "ammonitico rosso;" but judging from the overlying position of the nummulites in the south of Italy, as well as in the north, it is probable that the great mass or lower portion of alberese limestones is, I repeat, really cretaceous. See a translation of Prof. Savi's Memoir in Quart. Journ. Geol. Soc. vol. iii. part ii. p. 1–10.

ash-colour, on which Flaminius was defeated by Hannibal, are scarcely to be distinguished from those on which Caractacus made his last stand against the Romans, although the one is either eocene or younger cretaceous and the other Silurian palæozoic! I make this remark both as a fair excuse for the older geologists, who in a region so void of fossils had considered this macigno to be an ancient "greywacke"; and still more as a reason why the latter word should never more be used, except in a mineralogical sense. This solid macigno of the Thrasymene and Perugia, graduating upwards into thinner courses with flaglike calcareous bands, is surmounted by the pebbly, sandy, and marly accumulations on which Perugia is built*.

The whole of the western edge of the Apennines from Foligno to Rome is void of macigno, and the grand trough or basin, between that Apennine escarpment on the east and the ridges of secondary limestone of the Siennese and Roman Maremma on the west, is exclusively occupied by volcanic and tertiary deposits, through which a few islets or outliers of Apennine limestone, such as Mount Soracte, rear their heads as you approach to Rome. But to the south of Narni and in the Sabine mountains east of Rome, where the limestones are manifestly cretaceous, we again meet with overlying nummulitic rocks and macigno,—not indeed on the external or western face of the chain at Tivoli and Palestrina, but between those places and Subiaco. The chief limestone of this tract, even when in that state of marble called "Occhio di Pavone," has been found to contain hippurites. In traversing the chain from Palestrina to Subiaco, I perceived, that whilst it presented a broken and often abrupt escarpment to the plain of the Campagna, the hippurite limestone, when followed across its strike or to the east, soon folds over in rapid undulations accompanied by great fractures, and at Olévano is surmounted by an impure sandy limestone charged with nummulites and pectens. The whole calcareous series then plunges under troughs filled with macigno sandstone, precisely similar to that of Tuscany, and which, though weathering externally to rusty yellow and dirty ash-colour, is, when quarried into, the same dull bluish grey psammite with minute grains of black schist, so well known as the building-stone of Florence. These macigno beds are occasionally vertical, and often so broken and squeezed up between the older limestones (with a strike from S.S.E. to N.N.W.), that persons unaccustomed to their relations elsewhere might well be induced to suppose that they underlie the older rocks. At Rojati, however, which stands on a fine thick-bedded macigno with alternating layers of shale, that dips away at a slight angle (this place being near the centre of a trough), the rock passes downwards into the same sandy and siliceous limestones which form the summit of the picturesque cretaceous hill of Olévano. Again, at Subiaco (see fig. 35), the church of Maria della Valle is built on an inclined, nodular, grey macigno with soft partings, which, covered by a mass of unconformable and horizontal tertiary conglomerate, passes down-

* An accident which injured one of my legs prevented my exploring the hilly tracts east of Perugia and Assisi. But I could hear of no nummulites in the environs, and the Museum of the University does not contain them.

wards into the upper beds of the great cretaceous limestones, of which all the surrounding mountains are composed, and in the grottoes of which St. Benedict established his famous monastery. Here, as at Olévano, the beds between the solid hippuritic limestones and the macigno are sandy or siliceous, dirty or yellowish white limestones,

Fig. 35.

Sabine Hills.

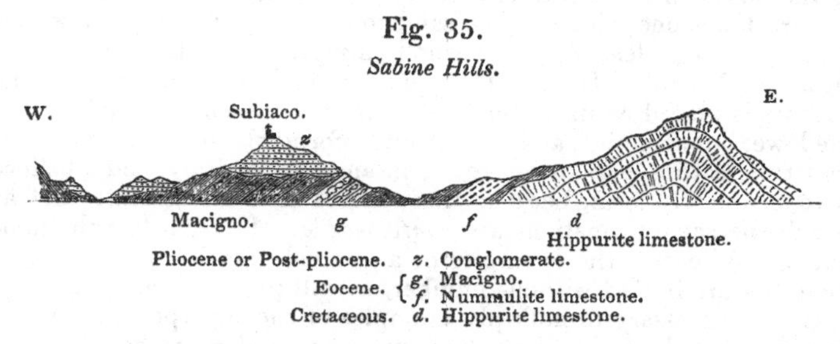

Pliocene or Post-pliocene. *z.* Conglomerate.
Eocene. { *g.* Macigno. *f.* Nummulite limestone.
Cretaceous. *d.* Hippurite limestone.

with nummulites and pectens. Near Agosta, lower down the valley of the Teverone, there are extensive quarries of this macigno where the rock, being deeply cut into, is blue-hearted, of concretionary forms on the great scale, and quite undistinguishable from the "Pietra forte" of Florence. The strata dip slightly across the valley and appear to plunge under the massive limestone cliffs of Agosta, but this appearance is fallacious, and is simply the result of one of the numerous faults of the chain; for the macigno is inclined at 10° or 12° only, and the secondary limestone plunges 45°. It is, I apprehend, from such examples that the supposed intercalation of the overlying macigno with the secondary limestones has been supposed to exist.

Judging from the section and brief description by MM. Alessandro Spada and Orsini* of the rocks between the watershed of the Apennines and the Adriatic in the parallel of Ascoli and the Tronto river, it would appear that there is there a much greater simplicity of structure than on the western side of the axis. This symmetrical disposition may be accounted for by the absence of those eruptive and volcanic rocks which are so abundant along the western slopes of the chain. Although I was prevented from visiting the Adriatic shores by the political state of the country, I cannot refer to the sections of Spada and Orsini without suggesting that one essential phænomenon of that region is in accordance with my own views. Their diagram shows a concordant passage from the limestone called "majolica" into overlying limestone with nummulites, and thence upwards through grey impure limestones with fucoids into macigno. Now whether this majolica be, as I suspect, neocomian and not jurassic (as they believe), and whether there be or be not any representative of the white chalk, we have clearly an ascending succession, in which the *macigno is the highest mass, and is overlaid only by tertiary miocene strata with gypsum.* The question after all is, what are the fossils which are there associated with true nummulites? and from

* Bull. Soc. Géol. Fr. 2 Ser. t. ii. p. 408, 1845.

all I could learn, they are there similar to those which have been described from other places. In other words, *they are not cretaceous*, but form the peculiar group in question. In this case, therefore, we have simply the well-known Alpine order of nummulite limestones, sometimes overlying neocomian and sometimes resting upon upper cretaceous rocks, and surmounted by vast masses of "flysch," *i. e.* of impure limestone with fucoids, sandstones, &c. There is, however, in this section a feature which wholly disagrees with the physical relations of Northern Italy. The gypseous molasse or miocene of the authors is placed as an unconformable mass between the macigno and the lower subapennine, and equally unconformable to both. I confess that this feature is unknown to me in any part of Italy, and I believe it to be merely local, because the authors themselves state that at Ascoli the same formations are *conformable*. In regard to the miocene and pliocene, the examples of a gradual transition from one to the other are indeed without number, as will presently be noted.

It is unnecessary to multiply examples of the superposition of the chief mass of the macigno to the youngest secondary rocks or hippuritic limestone of Central Italy. In following the upper road from Rome to Naples, any one may rapidly satisfy himself on this point at Ferentino, on the north side of which masses of inclined macigno with finely laminated sandy marls, dip away from a boss of the scaglia limestone and pass under all the younger and unconformable tertiary series of the Campagna. The masses of macigno on the south-western face of the great promontory of Sorrento, which forms the south side of the bay of Naples, also overlie cretaceous or hippuritic limestones, and the same order is seen in many tracts.

It is probable that the best fossiliferous exposition in the kingdom of Naples of both the upper cretaceous and the true nummulitic eocene, is exhibited in the grand Adriatic promontory of Monte Gargano, which it was my full intention to have visited, had the recent political troubles not prevented me. The late Professor Pilla is perhaps the only geologist who has examined and described it. But unluckily at the period of his visit he was not so well versed in stratigraphical geology and organic remains as he subsequently became, and I know from himself that he intended to revise his sections of that great headland. In default of a personal visit I was gratified to find, in the Royal Mineralogical Museum of Naples, so ably directed by Professor Scacchi*, a very illustrative series of the rocks and fossils of this Monte Gargano, the inspection of which left no doubt whatever in my mind, that the order of succession is there the same as that which I have witnessed in the Venetian Alps, the Papal States and other districts. The oldest rock is evidently the compact and hard white limestone with flints, and containing five species of Hippurites, besides Ammonites and Nerinææ. Then follow other beds, in which it would be presumptuous in me to attempt to decide the exact order. One of these is a red breccia; another is a peculiar

* Professor Scacchi himself intends soon to visit Monte Gargano and publish a detailed description of the order of the strata and their fossils. A full account of this promontory will form a beautiful monograph.

coralline, cavernous limestone with Pectens, Volutes, Olivæ, Dentalia, a large Fusus, Terebellum, &c.; a third contains Balani and Turbinoliæ. Now, the white limestone associated with this group is lithologically a true "calcaire grossier" loaded with nummulites. Among these nummulites, whatever may be their names, M. d'Archiac, who has examined them at my request, has declared that the four species to which he assigns them, all exist in the Lower Pyrenees. One of these, the *N. lævigata*, Lamk., occurs also in the London clay at Bracklesham, in the lower tertiary of Belgium, and in the Vicentine; and another is the form so very common in the Alps, whether it be termed *N. planospira* (Boub.) or *N. assilinoides* (Rüt.); whilst a fourth is the *N. granulosa* (D'Arch.) of Dax and the Pyrenees. These coincidences leave no doubt in my mind as to the age of the beds. I may also add, that in no one of the numerous rock-specimens I examined, is there an example of a nummulite occurring in the same fragment as the hippurites; and in fact there is also a clear lithological distinction between the hard, compact and flinty, white hippurite limestone, and the equally white but coarse granular limestone with nummulites.

The collections of Monte Gargano present, indeed, other fossils of a much younger series in a calcareous tuff, and as among these are the *Pecten latissimus, Panopæa Faujasi,* and other well-known subapennine shells, the existence there of pliocene deposits, as along other parts of the shores of the Mediterranean and Adriatic, is clearly marked.

On the Miocene, Pliocene, and Younger Tertiary Deposits of Italy.

The existence of deposits of miocene and pliocene age in the north of Italy has been long established; but geologists have not sufficiently directed attention to those sections in the peninsula, which best indicate transitions from the one group to the other. In the first place, therefore, I will endeavour to show, how in the north of Italy the oldest miocene, if not partly eocene, gradually inosculates and passes up into the overlying subapennine strata*. I have indeed already to some extent illustrated this point in the sections of Bassano and Asolo (p. 223), and have said that M. de Zigno will soon have collected, identified, and published the fossils which there lie in strata intercalated between the nummulitic eocene beneath and the subapennine marls and conglomerates above them.

The region to which I first invite attention, as exhibiting an uninterrupted succession from the top of the eocene or bottom of the miocene, through a full development of the latter up into the most copious accumulations of subapennine or pliocene, is that group of hills called the Monferrato, which ranges itself in a horse-shoe form, as defined by the course of the Po, between Turin on the west and Alexandria on the east. In the great plain occupied by coarse drift

* The reader who may desire to see the extent to which my observations and conclusions agree with those of other authors, may consult Pareto's Liguria Marittima, Pilla's Geologia, Philippi, &c.

which lies between the northern flexure of the Po, and the wall of crystalline and eruptive rocks which form the edge of the Alps, a great group of true cretaceous and older eocene formations may have formerly existed; but of these no traces now remain. The oldest rocks visible are those which, on the right bank of the Po, range from Gassino by Casal Borgone and the hills south of Verrua; in other words, they form the northern limit or escarpment of the Monferrato. Amid the sandy marls which there abound, certain peculiar and mottled limestones protrude, either in vertical or highly inclined positions, which, because they seem at one spot to have a certain direction and contain nummulites, have been described as *cretaceous* by M. Collegno* and by M. Sismonda. Since those authors wrote, the Marquis Pareto has in my opinion taken the true view of the subject, and has considered these beds to be tertiary and intimately associated with the miocene of the Superga.

Accompanied by the two excellent palæontologists of Turin, Dr. E. Sismonda and M. Bellardi, I made a transverse section of the ridge to the east of the ground which I had examined twenty years ago in company with Sir C. Lyell; and a simple description of this section will I hope set at rest the question as to the age of the lowest visible rocks in these ridges, and also indicate a gradual transition upwards into younger tertiary strata. In ascending the hill-side from the banks of the Po near Gassino, I found that micaceous marls dip south, and afterwards become vertical and graduate into calcareous sandstone locally called molasse. After passing over a slope obscured by vegetation, other sandy marls are seen to reappear at a higher level in vertical positions, and enclosing an equally vertical band of mottled, small concretionary limestone with nummulites (*g* of fig. 36), which strikes N.N.E. and S.S.W. A short interval is occupied by strata of green-grained marlstone with some imperfect minute plants, when there follows another and much stronger band of about 12 feet thick of similarly mottled, subconcretionary, blue-hearted limestone, which, striking from W.S.W. to E.N.E., plunges to the S.S.E. at an angle of from 50° to 60°, and is covered by impure limestone or calcareous sandstone. Ascending over a few undulating mounds of marl and sandstone (the site of old quarries where the limestone has been extracted), the same band of mottled limestone is met with, dipping at a high angle to the north.

To decide upon the age of this limestone (*g* of fig. 36), as M. Collegno has attempted to do, by noting the direction of any one of its broken masses, seems to me futile. Suffice it to say, that the chief

* Mémoires de la Soc. Géol. de Fr. vol. ii. p. 203. M. Collegno has also coloured in as cretaceous the whole of the eastern end of the Monferrato from Verrua to Casale, for which he has no better authority than the occasional reappearance of the Gassino limestones, sometimes with, sometimes without nummulites. I regret much that I had not time to explore the eastern part of the Monferrato, in which M. Collegno lays down a much broader mass of what he calls cretaceous; for judging from all other analogies, and even from what he writes himself, I have little doubt that downward passages there exist into the true eocene or equivalent of the flysch of the Alps; such rocks being, in fact, a part of the cretaceous group of that author.

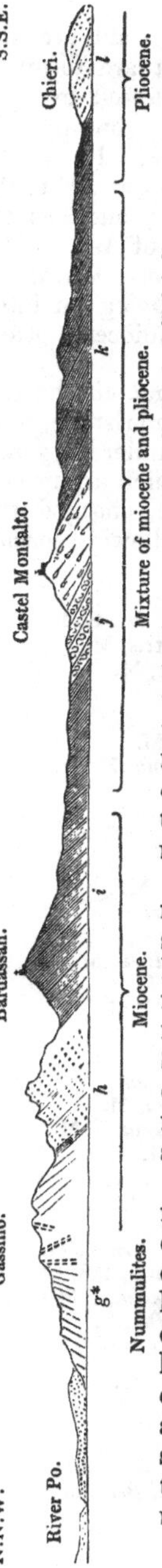

Fig. 36.

N.N.W. River Po. Gassino. Bardassan. Castel Montalto. Chieri. S.S.E.

g* h i j k l

Nummulites. Miocene. Mixture of miocene and pliocene. Pliocene.

mass of marls and sands strikes W.S.W. and E.N.E., and that such is also the strike of the principal mass of the mottled nummulitic limestone, and of the grand masses of conglomerate, sand and marl which constitute the higher and highly inclined miocene strata in the ridges of the Superga and Monferrato.

The nummulitic limestone (g^*) may be considered as an irregular axis, which throws off partially a few younger and broken beds to the north, that form the gypseous and other hillocks on the banks of the Po, and pass under the pliocene strata of Verrua and Crescentino; whilst it is manifestly succeeded on the south by the whole ascending series of the Superga and Monferrato.

It is also equally clear that this nummulitic rock is truly tertiary, if we judge from the other fossils associated with the *Nummulina placentula* (Desh.). Thus, the terebratula on which Collegno reckoned as a proof of cretaceous age, is now known to be a common species of the older tertiary of this tract. In this very limestone we found the great oyster, *O. gigantea*, fragments of pectens and corals, and above all the tooth of a fish (*Oxyrhina Desori*, Ag.), all well-known tertiary forms. Again, in the sandy beds, absolutely resting on those dislocated limestones, the *Pecten Burdigalensis* with Pectunculi and Turbinoliæ occur, and there is therefore no sort of doubt of the age of the rock.

I was much struck with the resemblance of the mottled, concretionary and coralline limestone of Gassino to the rock of Castel-Cucco between Asolo and Possagno, which is also the uppermost limit of nummulites (p. 222). I therefore consider, that in connecting the nummulitic base of the section of the Superga with a well-known band in the clear succession on the flanks of the Venetian Alps, I establish a connecting link between the eocene deposits of the Alps and the miocene of northern Italy.

In traversing southwards to Bardassan, across the ridges of conglomerate, both coarse and fine (h), which occupy the chief summits, and separate the valleys excavated in the softer marl or sandy shale, I had little to observe that is not already known; for these elevations and depressions are the direct eastern prolongations of the miocene of the Superga; but in descending from the hills to the south by Castel Montalto to the plain of Chieri, the development of the strata and the gradual change from the miocene into the pliocene type is too remarkable not to be specially noticed. On the south slope of the hill of Bardassan the coarse conglomerates have entirely disappeared, and are succeeded by sand, marlstone and marl (*i*), in which are forms of Cerithium, Cardium, Vermetus, &c. These are followed by sandstone (*j*) and impure calcareous concretions, which as you approach Castel Montalto alternate with finely laminated shale. In all this space (not less than two English miles), the strata dip at 20° to 25° to the south, and certain subapennine shells occur, such as the *Lucina Astensis* (a well-known species of the hills of Asti, which we found near Bardassan), associated with the *Cleodora obtusa* and *Ringicula Bonelli* of the Superga; thus already proving an intermixture of species which authors may cite, some as miocene, others as pliocene.

Pursuing the ascending section, and leaving behind all the true miocene sandstones, we come into marls, shale and marlstone, often white, which are overlaid by other marls descending under the yellow sands of Chieri. These last-mentioned marls (*k*), which are usually considered subapennine and are covered by sands (*l*), charged with the youngest known shells of the group, contain the following species, as named for me by Dr. E. Sismonda:—

Turbinolia duodecimcostata, *Goldf.*, P.
Terebratula DeBuchi, *Mich.*, M.
Ostrea cochlear, *Poli*, P.M.
Pecten cristatus, *Bronn*, P.
Nucula concava, *Bronn.*
  interrupta, *Nyst.*
  Nicobarica, *Lk.*, P.
  placentina, *Lk.*, P.M.
  rostrata, *Lk.*
Limopsis aurita, *Sassi*, M.
Arca diluvii, *Lk.*, M.
Lucina Astensis, *Bon.*, M.
  spuria, *Desh.*, P.M.
  transversa, *Bronn*, P.
Venus alternans, *E. Sismd.*, P.
Erycina complanata, *Récl.*
Bornia seminulum, *Phil.*
Dentalium circinnatum, *Sow.*
  coarctatum, *Lk.*, M.
  fossile, *Linn.*, M.
  inæquale, *Bronn*, M.
  pseudo-entalis, *Lk.*, M.
  rectum, *Linn.*, M.
  sexangulare, *Lk.*, M.
Cerithium vulgatum, *Brug.*, P.M.
  inflatum, *Bell.*, M.
Nassa costulata, *Ren.*, M.
  semistriata, *Br.*, P.
  serrata, *Br.*, P.
Buccinum polygonum, *Br.*, P.M.
Cassidaria echinophora, *Lk.*, P.M.
Cassis texta, *Bronn*, P.M.
  variabilis, *Bell. et Mich.*, M.
Cancellaria Bonelli, *Bell.*, M.
  calcarata, *Br.*
  lyrata, *Br.*, M.
  mitræformis, *Br.*
Cancellaria varicosa, *Br.*, P.M.
Pleurotoma brevirostris, M.
  cataphracta, *Br.*, M.
  Coquandi, *Bell.*
  denticula, *Bast.*, M.
  dimidiata, *Br.*, M.
  intermedia, *Bronn*, M.
  intorta, *Br.*, M.
  monilis, *Br.*, M.
  obtusangula, *Br.*, M.
  rotata, *Br.*, P.M.
  turricula, *Br.*, P.M.
  turritelloides, *Bell.*
  Rochettæ, *Bell.*
Raphitoma harpula, *Bell.*, M.
  plicatella, *Bell.*, M.
  vulpecula, *Bell.*, P.M.
  columnæ, *Bell.*
  textilis, *Bell.*, M.
Ficula ficoides, *E. Sismd.*
Fusus aduncus, *Bronn*, M.
  angulosus, *E. Sismd.*, P.M.
  crispus, *Bors.*, M.
  lamellosus, *Bors.*, M.
  longiroster, *Br.*, M.
  mitræformis, *Br.*, M.
Triton Apenninicum, *Bronn*, M.
Ranella marginata, *Sow.*, P.M.
  reticularis, *E. Sismd.*, P.M.
Murex craticulatus, *Br.*, P.
  funiculosus, *Bors.*
  fusulus, *Br.*, M.
  Lassaignei, *Grat.*, M.
  polymorphus, *Br.*, P.M.
  spinicosta, *Bronn*, M.
Typhis fistulosus, *Mich.*, M.
Columbella nassoides, *Bell.*, M.

Columbella thiara, *Bon.*, M.
 turgidula, *Bell.*, M.
Mitra scrobiculata, *Br.*, M.
 pyramidella, *Br.*
Conus antediluvianus, *Brug.*, M.
 bisulcatus, *Bell. et Mich.*, P.
 Brocchii, *Bronn*, P.
Chenopus pes-graculi, *Phil.*, M.
Turbo fimbriatus, *Bronn*, M.
Solarium moniliferum, *Bronn.*
Phorus testigerus, *Bronn*, M.
Nerita proteus, *Bon.*, M.
Natica helicina, *Br.*, P.M.
 pseudo-epiglottina, *E. Sismd.*, M.
 millepunctata, *L.*, P.M.
Ringicula Bonelli, *Desh.*, M.
 buccinea, *Desh.*, M.
 striata, *E. Sismd.*, M.
Turritella subangulata, *Br.*, M.
 varicosa, *Br.*
Bulla uniplicata, *Bell.*, M.

As the ascending series, in which intermixture takes place, is of considerable dimensions, and as even close to Chieri we still meet with a great number of Superga species, it is evident that a considerable thickness of beds may be classed either as miocene or pliocene, according to the forms which the observer may happen to meet with. Amidst the species collected from these blue marls, which are geologically subapennine (Castelnuovo and Pino), those marked M. exist in the miocene of the Superga; those marked P. are exclusively pliocene; and the individuals with the affix P.M. are common to the Superga and the true pliocene. Out of the 95 species, then, found in this one zone of blue marl, 16 are peculiar to it, 52 are known in the miocene, 10 in the pliocene, and 17 are common to the two formations.

The citation of this important fact teaches us, that the more closely the artificial limits of what geologists call formations are worked out, the more impossible will it be to draw fixed lines between natural groups of strata which, like these, have succeeded to each other without physical disturbances. At all events, wherever the different members of the same system so graduate into each other stratigraphically, mineralogically, and zoologically, the tints of colour by which they are characterized in a map should also be blended along such mixed frontiers.

In passing from the sandy beds in question by Castel Montalto to Pino and Chieri (the angle of inclination diminishing as we recede from the higher ground), the masses of which we have been speaking are conformably overlaid by a great thickness of yellow sands with some inosculating marls, which constitute the true subapennine beds of the Astesan, so well known to geologists through the works of Brocchi and others. In these uppermost beds nearly all traces of anything purely miocene have disappeared, and we are immersed in that same type of shells with *Panopæa Faujasi*, &c. which at St. Gallen and other places characterizes the marine molasse of Switzerland (see *antè*, p. 230)*. I specially, then, caution geologists against employing that term in a sense which is to convey an idea of age, for as used at Turin the word *molasse* is exclusively applied to the strata of true miocene age, whilst in Switzerland the greater part of it is pliocene. Again, the pliocene deposits in Switzerland are hard sand-

* This miocene is laid down in Collegno's map, but in my opinion, as above explained, a great error prevails in that part of it which represents the eastern portion of the Monferrato as cretaceous.

stones and conglomerates, whilst in Italy they are soft marls and sands.

The true pliocene deposits of Asti occupy a broad trough, watered by the rivers Tanaro and Bormida, on the southern side of which rise up those micaceous and often greenish sandstones of miocene age, so largely displayed in Piedmont. On a former occasion (1828) I traversed a large breadth of these between Savona and Acqui in the company of Sir C. Lyell, and in my last visit I examined them in travelling to Genoa from Alexandria. Between Gavi and Arquata, they have all the characters of a regenerated macigno, and at Serravalle and Ligurosa rise up from beneath the subapennine marls and sands in highly inclined sandstones and marls, underlaid by powerful bands of conglomerate that dip 40° to the N. or N.N.W. In this manner we reach the opposite or southern side of the great tertiary trough of the Astesan, and are again in the equivalents of the Superga conglomerate. I could discover, however, no course of underlying nummulitic limestone similar to that of Gassino. At the same time it must be stated, that the system of macigno and alberese, which is considered by Pareto to be the equivalent of the nummulitic group (?), succeeds near Ronco, dipping at a high angle under the whole of the conglomerate and miocene series of Piedmont. I cannot positively say whether these underlying beds of flagstone and macigno on the south side of the basin are conformable to the overlying miocene series, but in a rapid survey they seemed to me to be so, and also to be in a much less crystallized and altered state than in the environs of Genoa.

The miocene of Piedmont contains the coal deposits of Caddibuona *, so long known and so often described, on account of the remains of the Anthracotheria associated with lacustrine and fluviatile fossils; and as we travel down into the peninsula similar examples are met with. The interstratification of freshwater or estuary beds (containing Melanopsis, Melania and Neritina) with marine tertiary strata, has been pointed out as occurring in several parts of the north of Italy by the Marquis Pareto †. Near Siena, as will presently appear, such beds manifestly inosculate with the upper strata of subapennine age; whilst in the environs of Tortona they seem (judging from that author's section and description) to lie low in what must be defined as the true pliocene formation. The fact, however, is, that as some of the acknowledged miocene strata of the peninsula are of terrestrial and freshwater character (Caddibuona, Monte Massi and Monte Bamboli in the Maremma, &c.), there can be little doubt, that the more observation is extended, the more evidences shall we find of such local freshwater intercalations throughout the tertiary series in many parts of Italy.

* I visited this place with Sir C. Lyell, in passing from Savona to Acqui. Its powerful conglomerates are possibly of the same age as those of the Superga (see Lyell's Principles of Geology, 1st edition, vol. iii. p. 221, woodcut No. 55, and 4th ed. id. vol. iv. p. 152).

† See the Marquis Pareto's memoir, read at the Scientific Congress of Turin, 1844, entitled "Sopra alcune alternative di strati marini e fluviatili nei terreni di sedimento superiore dei Colli Subapennini."

At Canniparola, on the west side of the valley of the Magra near Sarzana, coal supposed to be of miocene age has been worked in several seams, and is associated with many plants of dicotyledonous structure. These works are now abandoned. On exploring the natural outcrop of the mineral, on the sides of the torrent called La Girona, I perceived that the coal seams subordinate to shale with plants, reposed, in highly inclined strata dipping to the west, on shale with calcareous nodules. The latter passes downwards into a soft macigno or finely micaceous sandstone, and from that into dark-coloured and party-coloured schists and marlstone with conchoidal fracture. Below these there appeared to me to be a transition into true hard and older macigno with white veins, the inclination increasing to verticality with the slope of the mountain sides. These beds being partially dislocated, strike both a little to the west and a little to the east of north; but the main direction is north and by west, and south and by east, or parallel to the chief ridge of the Apuan Alps, from which they are separated by a vast mass of underlying macigno. On the other hand, the coal strata are surmounted by ochreous sandy conglomerates, which being further removed from the axis of elevation, dip down to the vale of the Massa at a less inclination, and are lost in alluvial accumulations.

A traverse of the Apennines from Bologna to Florence exhibits, on the flanks of the chain near the former city, blue marl and sand of subapennine age, reposing on micaceous sandstone.

These masses, thus exposed in a low anticlinal on this outer parallel, are better seen on the high road as you pass by Pianura to Lojano, arranged in an elevated trough, near the north-western side of which courses of lignite (1) are surmounted by nodular strata and shelly blue marls, and these by the sands and white marls on which Pianura stands. Above these come other sandy marls with large nodules of dark grey, micaceous, ferruginous marlstone, in which I found many Cardia, Pectunculi, Nuculæ and Venericardiæ. These shelly beds, overlaid by a vast thickness of blue marl (2), and covered by yellow and white conglomerates and sands (3), are clearly the subapennine group of Brocchi, which, after dipping for a certain distance to the west, are bent up in a trough. From the summits of the conglomerate hills near Lojano, the dip being reversed, or to the N.E., the subapennine group is supported on the other side of the basin by nodular strata, together with a system of soft micaceous sandstones and pebbly conglomerates of considerable thickness, which alternate with certain shaly marls and greenish sandstones. These lower, undulating sandstones and conglomerates with marls, &c. most clearly represent the Superga series (*h* of fig. 36), and are of miocene age. In the construction, therefore, of a detailed geological map, this portion of the east flank of the Apennines might be shown to exhibit two axes or undulations of miocenic sandstones and conglomerates, troughing between them a mass of true subapennine beds, and again throwing off the upper beds towards the low country. At Lojano, the second post from Bologna (as it appeared to me in a rapid survey), the miocene conglomerates are cut off by a longitudinal fault from the macigno,

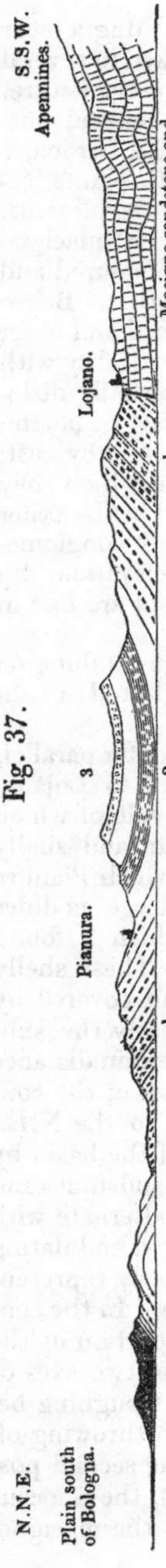

Fig. 37.

N.N.E. Plain south of Bologna. Pianura. Lojano. Apennines. S.S.W.

Macigno sandstone and limestone (Alberese).

3. Yellow sands and conglomerates. } Subapennine or Pliocene.
2. Blue marl.
1. Miocene conglomerates and sands, &c.

which succeeds to the west: for the latter is only slightly inclined to the east, and is a hard, micaceous, true macigno sandstone, which, flaglike near the surface, passes down into thick beds. This rock, receiving its peculiar tint from numerous minute fragments of black schist, is undistinguishable from the macigno alpin or flysch of the Alps. It occurs, in fact, on the western edge of those great undulations of alberese, and other limestones, which, perforated by serpentine at Monte Berici and Sasso di Castro near Coviliajo, form the chief mass of that group already spoken of, whose geological equivalents in the absence of fossils it is so difficult to define.

Thus, whilst the Bolognese Apennines expose an intimate connection between the miocene and pliocene groups, they afford, as far as I saw, no indications of an unbroken succession from the macigno to the overlying miocene. It appeared, indeed, to me, that in descending towards Florence by Campo Santo and Crespiano, a conglomerate (probably miocene) was there also adherent to the sides of the mountains of older date; but in that portion of Tuscany the union between the miocene and pliocene, as above described, is wanting.

The picturesque hills around Lari, on the south side of the Arno near Pisa, which I visited with Professor Pilla, are for the most part composed of subapennine blue marl, loaded with shells and covered by yellow sands; the *Ostrea hippopus*, *Pecten laticostatus*, and large *Panopæa Faujasi* lying about in abundance. The villages stand on insulated points of the overlying sands or sandy marls, the remnants of former great denudations, all the strata being horizontal. These elevated sandy and loamy points are rich and fertile, whilst the denuded argillaceous marls of the valleys are sterile;—physical features which so prevail in vast tertiary tracts throughout Italy, that the agricultural characters alone are there sufficient to indicate the age of the strata. Near Casciano, however, to the south of Lari, as well seen in the quarries of Sta Frediana, other and lower sandstones of harder character, rise out abruptly from beneath the subapennine cover, and form broken and undulating domes. These beds contain highly ornamented small Echini, small Ostreæ, and other shells, together with fishes' teeth and palates, unknown in the overlying formation. The strata are specially characterized by as oft, calcareous yellow sandstone, arranged in large concretionary shapes, which here and there passes into limestone and calc grit, but in many parts disintegrates into fine yellow sand, in which caverns have been excavated. It is this rock which has afforded the numerous Lenticulites and other foraminifera described in the works of Soldani and Targioni-Tozzetti. They are accompanied by a very minute Terebratula, to which M. Pilla particularly directed my attention, and which at first sight had much the aspect of forms known only in palæozoic rocks*.

There can be no doubt that the foraminiferous rock of Sta Frediana is of miocene age, but as it has here been brought up through the subapennine strata, along one of those lines of fracture so common in the adjacent region of the Maremma, we naturally miss the links, stratigraphical and zoological, which connect the miocene and pliocene in the Monferrato of Turin and in the Lower Apennines of Bologna.

Further southward, and in entering the Tuscan Maremma, rocks of this miocene age re-occur, overlying the ridges of alberese and macigno which there rise up, and in one place, Botro di Laspa near Pomaja, *i. e.* in the direct road from Pisa to the Maremma, contain the same small Terebratula as at Frediana. I examined the flanks of the lateral valley through which that route passes, and where the miocene contains large masses of gypsum. Traversing the hills from Castellini to the copper-mines of Monte Catini†, I thence made an excursion into the heart of the Tuscan Maremma to explore the relations of the coal beds of that tract which have been so largely opened out, and would, doubtless, have been rendered useful, had not revolutionary agitation checked all public and private expenditure.

* See " Osservazioni sopra l' età della pietra lenticolare di Casciano nelle colline Pisane, di Leopoldo Pilla." In this notice, published after a joint examination which I made with him, Professor Pilla corrects a first sketch, in which he had considered this lenticular limestone as of subapennine age, and shows that at Sta Frediana and Parlascio it constituted islets or reefs of rock of miocene age in a sea of subapennine age. The corals, Lenticulites, Echini, Terebratulæ, &c., are supposed to be miocenic, whilst certain Ostreæ and Pectens are presumed to be pliocenic. It does not, however, appear that the latter are identical with known subapennine species. My lamented friend Professor Pilla had formed an opinion respecting the usual horizontality of subapennine strata, as contrasted with the inclination of all beds of miocenic age, in which I cannot participate. In this case I believe that the oldest tertiary of this part of the basin has been heaved up through the overlying strata on lines from north to south; and I cannot agree with him, that these older masses ever formed ancient islets, around which the younger were accumulated. On the contrary, I am convinced that here, as in the Monferrato, the whole submarine tertiary series was *originally* deposited successively and without a break.

† At Monte Catini, where I was hospitably received by Mr. Sloane, the intelligent proprietor of the mines, and in other places, I examined the serpentine, gabbro, and other eruptive or unstratified rocks, into the consideration of which I shall not enter in this memoir, the object of which points exclusively to sedimentary succession. The chief phænomena have been already described by Mr. W. Hamilton, Journ. Geol. Soc. London, vol. i. p. 291, and have been copiously dwelt upon by Savi, Pilla, and others.

Micaceous sandstone, which I believe to be of miocene age, with traces of stems of plants, appears in the conical hillock near Monte Catini; but the strata are there so dislocated in their relations to serpentine and gabbro, that no distinct order is visible. It was the belief of Professor Pilla, that much of the argillaceous and sterile marl of the deep denudations around Volterra, particularly the lower portion which contains large masses of gypsum and salt-springs, is also of miocene age. Of this there are no fossil proofs of which I could hear. It is, however, certain that the thick unfossiliferous marls are surmounted by others, and finally by yellow sands and sandstone, the "panchina," on which stands the noble ancient city of Volterra. These are true subapennine beds with many fossils; the tombs of the Necropolis being excavated in the sandy "panchina."

Pomerancia, to the south of Volterra, is placed on a high plateau of shelly tuff, which probably pertains to the upper portion of the pliocene, but the mountains to the east and south are macigno and alberese (possibly of cretaceous age?), with nuclei of still older rocks. Not now adverting to these rocks, or to the hot springs issuing through them which afford the boracic acid*, I will briefly notice the coal deposits of Monte Bamboli and Monte Massi, which lie still further to the southward. These deposits are described by Savi and Pilla, and the coal has been analysed by Matteucci. For my own part, I consider them to be of about the same age (miocene) as the coal of Caddibuona in Piedmont and of Fuveau near Toulon in the south of France†.

At Monte Bamboli the coal-seams, varying from eighteen inches to five feet and inclined about 30°, rest on earthy and broken schists, which are so nearly in contact with the surface of the so-called alberese of this tract, that I could scarcely divest myself of the idea that the one had succeeded conformably to the other; but although the upper schists or galestri of the alberese appeared to graduate into the gritty schist and the latter into the coal-seams, the whole dipping to the W.S.W., I was subsequently led to believe, from the sections at Monte Massi, that the apparent conformity is accidental. The coal, of which there are here two courses, is interlaced with a band of an earthy, shelly, freshwater limestone with mytili; and the surface of the coal-seams, in which plants and shells occur in a schist or "bat," graduates up into a considerable thickness of a thin, flat-bedded, sandy, impure limestone, which is followed by an indurated clay rock, and the latter by a coarse conglomerate. Whilst the coal-beds dip 30° S.W. in the shaft west of the works, they may be observed in the side of the torrent north of the engine to roll over to the N.N.E., and in one spot they dip 70°. In these dislocations

* In reference to the various, intensely hot springs which afford the boracic salt, I will only here say, that they seemed to me to issue from fissures having a direction from north and by west to south and by east, and that these parallel linear outbursts thus seem fairly to represent the last remnants of that grand subterranean evolution of heat which in former ages has so affected all this range of the Maremma.

† See description of this coal-field by Sir C. Lyell and myself, Proc. Geol. Soc. vol. i. p. 150.

the coal seems to follow all the accidents and undulations of the subjacent, so-called alberese, limestone on which it rests. Besides mytili and plants, the tooth of a pachyderm has been found in this tertiary coal, which M. Pomel has named *Iotherium*.

The other accumulations of this age occupy broken troughs throughout the Tuscan and Roman Maremma, and those which I visited, lie to the south of the rocky village of Monte Massi, where three shafts have been opened and where the coal is much developed (see fig. 38). Eruptive rocks, chiefly of serpentinous character, occupy the summits, and on one of these the grotesque village of Monte

Fig. 38.

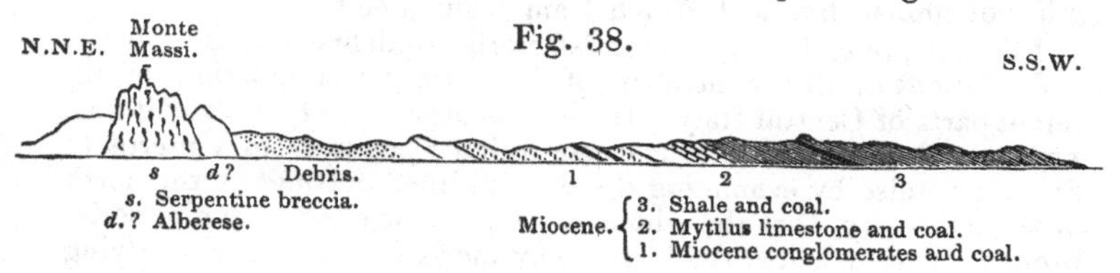

s. Serpentine breccia.
d.? Alberese.

Miocene. 3. Shale and coal. 2. Mytilus limestone and coal. 1. Miocene conglomerates and coal.

Massi is perched. The rock is here a serpentinous breccia, classed as "euphotide," which throws off vertical patches of alberese (*d*) on all sides. But instead of the short interval which occurs at Monte Bamboli between the surface of the alberese and the coal, we have, first, a conglomerate of alberese, secondly, a thick mass of grey stratified shale or clay, and thirdly, grits and small pebbly conglomerates, with fragments of serpentine (ophiolite grit of Savi), on the last-mentioned of which the lowest bed of coal reposes. This succession is obvious in descending from Monte Massi to the banks of the brook on the S.S.W., in which the natural outcrop of the lower coal is seen. Even this lower coal (1) is not considered the same as that of Monte Bamboli; for after ascending through shale and grit, &c., another seam occurs which is interlaced with and surmounted by the very same *mytilus limestone* as that of Monte Bamboli (2), the whole dipping away at about 25° to S.S.W. Then follows a considerable breadth of argillaceous shale, the angle of inclination decreasing as the beds advance into that broad valley which terminates in the mouth of the Ombrone at Grossetto. Subordinate to this shale and claystone is the third or great seam of coal (3), which is of considerable thickness, and into which I descended by the new pits. Many portions of the coal, whether judging from the eye or from its chemical analysis, differ little from the inferior but useful qualities of British combustible of the palæozoic age. Unluckily, however, both Monte Massi and Monte Bamboli are at some distance from the seaboard, and no rail- or tram-roads having been yet constructed, all the expenditure of the miners will be thrown away if public assistance be not given to them*. The statistical data of these coal tracts, the great heat experienced in the deep shafts at Monte Massi, the gases (not inflam-

* I was accompanied to these tracts by M. Caillaud, the principal director of the Leghorn Coal Company, and by M. Petiot, the intelligent French engineer who has directed the works.

mable) therein, and other points of great interest, must now be passed over.

If a geologist examined the district of Monte Bamboli only, he might form a conclusion that the beds with coal at once succeeded to the alberese limestone; but at Monte Massi he sees that one underlying conglomerate is formed out of that rock, and that another, derived from the serpentines associated with it, forms the absolute base and support of this capacious and very remarkable coal tract, which although referred to the miocene age and clearly subjacent to all the lower subapennine, has more the aspect of an old coal-field than any other of similar date with which I am acquainted*.

I do not pretend to be able satisfactorily to define the exact limits and relations of all the members of the tertiary accumulations of different parts of Central Italy. In the southern part of Tuscany it was, however, clear to me, in a traverse which I made from Volterra to Siena, and also by examining the deep railroad cuttings to the north of the latter city, that the whole of the pliocene or subapennine series properly so called, *i. e.* the blue shelly marls (1) and their overlying yellow sandstone (panchina) and conglomerate (2), are there surmounted by the freshwater limestone (3), which occupies plateaux between Monte Reggioni and Colle, and in a deep denudation at the latter place is seen to rest on shelly subapennine strata, as expressed in the opposite diagram (fig. 39). Near Castello St. Geminiano on the west, the yellow subapennine sands with shells rise out rapidly from beneath this tufaceous limestone with its Lymnææ, Planorbes and other shells, and at Monte Reggioni on the east a similar infraposition is equally clear. Towards Siena this freshwater formation becomes a massive travertine, and constitutes undulating hills of hard and tough cavernous rock; among the lower masses is a very coarse conglomerate with huge angular fragments of Apennine limestone, often two and three feet in diameter. The strata of reddish colours, which are cut through by the railroad towards the source of the Staggia, are evidently a portion of this same irregularly deposited and block travertine, the whole of which overlies the subapennine group. In this tract there are considerable fractures, and wedge-shaped masses of the shelly blue marl are here and there forced up against the overlying conglomerate and travertine.

The more detailed order of this district, which cannot, however, be expressed in a general woodcut, seemed to me to exhibit in a descending series beneath the vegetable soil, 1st, coarse alluvia; 2nd, finely laminated sandy loam; 3rd, lacustrine limestone with Lymnææ

* For details respecting these coal tracts of the Tuscan Maremma, see Professor Savi's work, "Sopra i carboni fossili dei terreni mioceni delle Maremme Toscane, Pisa, 1843." Among the fossils he cites bones, possibly belonging to carnivora, and teeth of rodentia, *Mytilus Brardi* (Brongn.), opercula of univalves, and imperfect casts which may belong to Buccinum, Fusus and Cardiaceæ. The characteristic plants are Palmacites, a Musacea termed Uraniophyllites by Prof. Pietro Savi, with leaves of various dicotyledons (oak, plane, elder, cornel), cones of pine, &c. Prof. Pilla has also described these tracts in a memoir entitled "Sopra il carbon fossile trovato in Maremma," Florence, 1843; and in a work called "Breve Cenno della richezza minerale delle Toscana," Pisa, 1845.

Fig. 39.

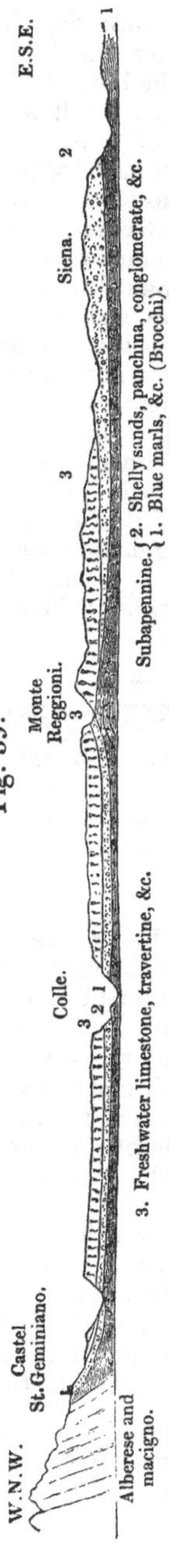

W.N.W. Castel St. Geminiano. Colle. Monte Reggioni. Siena. E.S.E.

Alberese and macigno. 3. Freshwater limestone, travertine, &c. Subapennine. { 2. Shelly sands, panchina, conglomerate, &c. 1. Blue marls, &c. (Brocchi).

and Planorbes, based upon and passing into a coarse travertine, with calcareous agglomerate and breccia; 4th, conglomerates of Apennine limestone, the surfaces of the rounded pebbles being often covered with Balani and Serpulæ, and associated with yellow sands containing large Ostreæ and Pectens, &c.; these graduate down into calciferous yellow sandstone, the "*panchina**" building-stone of the country, with concretions of calcareous grit, &c.; 5th, blue marls, which are much richer in shells in their upper parts only, where they graduate into the yellow sands.

The larger and lower portion of these marls is, indeed, throughout large tracts as sterile in organic remains as it is in its agricultural character. The desolate region between Siena and Radicofani is entirely composed of these naked, dull grey marls. On the other hand, the pebble beds and incoherent sandstones and marls on which Perugia stands, and in which her ancient Etruscan tombs were excavated, are probably of miocene age. At all events they resemble the Superga series in mineral aspect, repose upon macigno and alberese, and at Ficullo, between Perugia and Orvieto, are succeeded by unquestionable and very shelly subapennine strata.

In the volcanic regions extending from Radicofani to Rome, there are no evidences of any stratum older than the blue subapennine marl. To find the equivalents of miocene deposits in the southern parts of the Papal States, we must either travel eastwards into the valleys of the Apennines, or, passing the axis, explore the rich deposits of that age with plants and shells which await the geologist, who will work out the data of which M. Orsini and Count A. Spada have given a sketch.

Not intending now to enter upon the consideration of the volcanic dejections of the Papal or Neapolitan States†, I would however say a very few words to show the connection which exists between the subapennine strata that crop out at Rome and their associated rocks, and thus indicate generally the succession of geological phænomena in the environs of the ancient

* Although applied to rocks of pliocene age at Siena and Volterra, the term "panchina" is used at Leghorn in reference to a marine tufa or shore deposit, covered by red earth, which is younger, and probably of the same age as the overlying travertines (3) of Siena and Colle.

† Monsignore Medici Spada and Professor Ponzi prepared and presented to me a map of the volcanic dejections of Latium, which they regard as terrestrial, in contrast with the subaqueous formations of the Campagna. They have indeed published a general section, "Profilo teoretico dei terreni della Campagna di Roma."

mistress of the world. The oldest beds visible (and these only in deep denudations to the north and west of St. Peter's and the Vatican, and at the foot of the Monte Mario*) are the blue shelly marls or clay, largely excavated for brick-earth. These are followed by sandstone, occasionally calcareous ("panchina" of Tuscany), by yellow sands, and finally by pebble-beds, the materials of which have nearly all been derived from the Apennine limestone. This is the subapennine shelly group of Brocchi, and in reference to its fossils I may state, that although many forms are common to the lower and upper beds, which are so strikingly distinguished from each other in lithological aspect, there are certain species, such as the *Cleodora lanceolata* and *C. Vaticani*, which pertain to the inferior blue marls only, and are never found in the overlying yellow sands. It may also be mentioned, that here, as at Siena, the greatest number of shells are found in the beds of junction of the two divisions.

The order in which the blue marls (the oldest stratum of the district) are overlaid by yellow sandstones and pebble-beds, and the manner in which the latter were first associated with and next covered over by volcanic materials and then elevated into land, and what changes the surface subsequently underwent, I would attempt to explain in this general woodcut, fig. 40. On the flanks of Monte

Fig. 40.

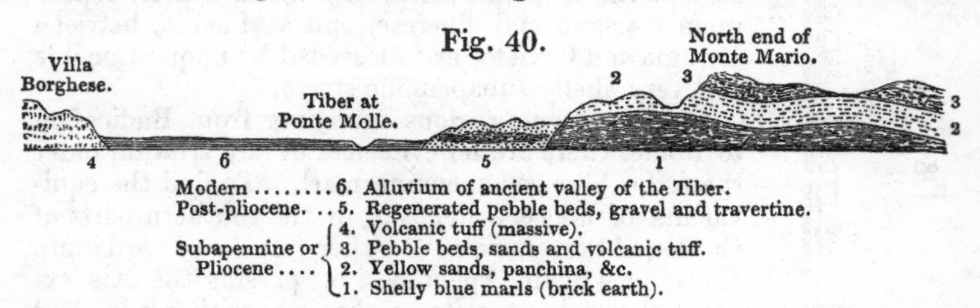

Modern .... 6. Alluvium of ancient valley of the Tiber.
Post-pliocene. .. 5. Regenerated pebble beds, gravel and travertine.
Subapennine or Pliocene .... { 4. Volcanic tuff (massive).
3. Pebble beds, sands and volcanic tuff.
2. Yellow sands, panchina, &c.
1. Shelly blue marls (brick earth).

Mario, good evidences exist of the superposition to the blue marls (1) of the yellow calciferous sandstones and sands (2). In mounting to the overlying pebbly beds (3) we see the first commencement of submarine volcanic action in dejections of finely laminated peperino and tuff, which are dovetailed into the uppermost of the subapennine strata. Then follow those tuffs, peperinos and other volcanic rocks of the Campagna, which were so extensively spread out under a former sea, and of which the hills of Rome and the Villa Borghese afford examples and varieties (4). These are, in fact, the submarine accumulations which terminated the subapennine period.

After such masses had been raised into land, and when the valley of the Tiber became in the first instance either a lake or a broad river, detrital accumulations were, it would appear, formed out of the materials both of the pliocene strata (1 & 2) and also of all the subaqueous volcanic dejections (3 & 4) which had overspread them. The materials of the ancient gravel above Ponte Molle decide this

* It would appear that a good many species of shells have been detected at Villa Madama, the Vatican and other localities since Brocchi wrote. It is not my province to enter into these details.

point. Such terrestrial deposits, antecedent however to our own æra, there form low hillocks of gravel and sand, including fragments of the submarine volcanic rocks (4), and also a band of travertine. It is in these accumulations that numerous remains of the quadrupeds which inhabited primæval Italy are found. Professor Ponzi has clearly distinguished* them from their congeners in the older period (3) or upper strata of the Subapennines. In that preceding period the *Elephas primigenius* (Blum.), *Hippopotamus major* (Blum.), *Rhinoceros leptorhinus* (Cuv.), *Equus fossilis* and *Cervus primigenius* were inhabitants of the adjacent Apennines, from which their bones, with much pebbly detritus, were washed down into the adjacent estuaries and bays of the sea, and mixed up with its dolphins and shells. When the estuary formations had been raised into land and formed the banks of the ancient or broad valley of the Tiber, other quadrupeds appeared, and if the bones of the older period be found added to more recent remains, the former are always in a rolled and water-worn condition.

Among the animals of the post-pliocene or quaternary deposits (5) whose remains have been detected in such hillocks as those at Ponte Molle, are *Ursus*, *Meles antediluvianus*, *Felis brevirostris*, *Sus scropha fossilis*, *Equus fossilis*, *E. asinus fossilis*, *Cervus primigenius*, *Bos priscus*, *Bos primigenius*, with aquatic birds, frogs, eels, &c.

From that epoch, so recent as respects geological history, but so remote as respects man, we are ushered into our own æra by finding in the more modern alluvia of the Tiber, but when that stream was much broader (6), the remains of creatures, such as the *Dama Romana*, the *Ovis aries* and *Capra ægragus*, which, though comparatively recent and having disappeared from the peninsula, are in this last deposit associated with the usual modern types, including the *Bos bubalus* (Linn.), which shows that the Buffalo is indigenous in Italy.

In reviewing the vibrations and changes of relation which the tertiary deposits of Italy have undergone, it appears that though in many districts there are dislocations which affect one group and not another, there are, on the other hand, sufficient examples of transition which unite them. In this manner we have seen instances where true eocene, as proved by organic remains, passes up into miocene beds equally upheaved and conformable to them (Bassano, Asolo); whilst in the southern parts of Tuscany and in the north of the Papal States remains are seen in masses, which though much less fossiliferous are presumed to be their equivalents. Some of the miocene coal deposits of Tuscany follow all the flexures and dislocations of the older rocks on which they rest. M. Coquand compares them with those of Aix in Provence and other spots well known to him, and finding that they contain the same characteristic plant, *Palmacites Lamanonis*, he has contended that they should even be classed as eocene or with the gypseous beds of the Paris basin. In synchronizing freshwater with marine deposits, where there is not a continuous succession of many strata, there is always considerable difficulty; but

* See the Atti della ottava riunione degli Scienziati Italiani Genova, pp. 679 *et seq.*

as these lignites are manifestly posterior to any stratum of the nummulitic series, which I regard as the eocene of Southern Europe, I must consider them to be of miocene age, though in some instances representing perhaps the lower beds of that division. In the full and consecutive marine series of the Monferrato, and in the Apennines of Bologna, no doubt can remain of a perfectly equable and conformable transition from miocene into pliocene. Even in the convulsed region of the Tuscan Maremma and its flanks, it is manifest that beds of miocene are surmounted by the whole series of the lower gypseous marls, which in their turn, though often highly inclined, pass up into true subapennine blue marl.

Some geologists have indeed endeavoured to distinguish the miocene from the pliocene tertiaries of Italy by the inclination of the strata in the one and their horizontality in the other. But this method is fallacious; for although the great shelly masses of pliocene age, which occupy broad valleys or large troughs, are necessarily more or less horizontal, wherever they are removed from centres of disturbance, there are numerous districts in which they are highly inclined. Thus, without going back to the sections of Bassano, the Monferrato, Bologna, &c., we see both the blue marl and yellow sands, which are so horizontal along the banks of the Elsa in Tuscany, dip at 35° to the east of Volterra, whilst they are followed downwards at Specchiajolo and Pignano, on the road to Colle, by gypseous marls, which are still more highly inclined as they approach a ridge of elevation. Again, where the basaltic cone of Radicofani perforates the tertiary trough of blue marls which lie between the ridges of Monte Amato on the west and Monte Cetona on the east, these young strata are singularly dislocated. Even without quitting the environs of Rome, the most perfect horizontality of the blue marls and overlying sandstones may be observed near St. Peter's and the Vatican; and yet in following the uppermost of these strata to the summits of Monte Mario or westwards towards Civita Vecchia, they are found to undulate so rapidly with local breaks, that sections made in two detached spots would show an apparent unconformity, when in fact all is one continuous series.

On the shores of Italy, as in the valleys of the Arno and the Tiber, there are many proofs of a succession of deposits similar to that which has been alluded to near Rome, ascending from the subapennine or pliocene æra into the period when all the sea shells found in the raised beaches are those of the present sea. On this point I will now only add, that the oscillation to which the coast has been subjected in the historic period, when the temple of Serapis in the bay of Puzzuoli was depressed about twenty-five feet below its present level and afterwards raised, was by no means a mere local subsidence, but one which affected the whole of the adjacent coast of Italy. For, on the seaward face of the promontory of Gaeta, which is a mass of subcrystalline or hippurite limestone, I satisfied myself of the accuracy of the observation of Pilla* and other Italian geologists, that pholades of existing species had eaten into the rock at about the same height

* Trattato de Geologia di L. Pilla, vol. i. p. 334. Pisa, 1847.

above the water at which the perforations of these animals are observable in the columns of the temple of Jupiter Serapis; whilst subsidences of ancient Roman buildings beneath the sea are apparent in many adjacent places.

But although we thus learn that such oscillations of the land have been in operation during the historic æra, who will venture to compare the operations which gradually elevated and depressed the coast of Italy a few feet, with those mighty forces which evolved the more ancient upheavals, fractures and inversion of the Alps and Apennines? By no amount of gradual intumescence and subsidence can we explain the grand phænomena of those mountains, and the geologist cannot examine them without admitting, that they stand forth as monuments of much more powerful causes than any of which there is a trace in the modern period.

Concluding Remarks.

In recurring to the chief object of this memoir—the recognition of Eocene deposits of large dimensions in the South of Europe—it is unnecessary that I should here enumerate all the authors who have considered the nummulitic rocks of the Alps and Italy to be of cretaceous or secondary age; it being enough to state that in the works of É. de Beaumont, Dufrénoy, Studer, Escher, and others, and in nearly all published maps and tabular views, they are still so classed. Having now entirely abandoned the opinion which I once entertained, that nummulites are common to the cretaceous and tertiary rocks of the Alps, as explained in the preceding memoir, I will endeavour to generalize the result. But first let me pass in review those authors who have recently thrown light upon this subject by their surveys in the south of France, where, in proceeding from our northern countries, we find the eocene formation beginning to assume its Alpine and Mediterranean aspect, and what I consider to be its great and normal type.

Our associate Mr. Pratt, who has so well illustrated the case of Biaritz at the north-western foot of the Pyrenees, believes, that the nummulitic and shelly strata there exposed are tertiary*; but whilst a great number of the fossils (56 species) are identical with forms of the Paris basin, he conceives that the strata are of somewhat older date than the eocene of the north of Europe. This opinion is probably to some extent correct, since a portion of the beds in question may represent that interval of time which is marked in England by the great disruption between the plastic clay and the chalk. In examining the fossils collected by Mr. Pratt, M. d'Archiac detected only three cretaceous forms in 108 species†, and of these, two are individuals, *Ostrea vesicularis* (Sow.) and *O. lateralis* (Nilss.), which are repeated in other tracts in the lower stage of the nummulitic formation.

In dividing the nummulitic group of the basin of the Adour into three stages, M. Delbos shows that its inferior member, containing

* Bull. Soc. Géol. Fr. 2 Ser. vol. ii. p. 185.

† Mem. Soc. Géol. Fr. 2 Ser. tom. ii. p. 191.

the *Ostrea vesicularis* (associated, however, with the tertiary species *Ostrea gigantea*, *Terebratula semistriata* and *Cancer quadrilobatus*), reposes on strata charged with *Inoceramus Lamarckii* and *Ananchytes ovatus*, which he believes to be the true representative of the white chalk of Paris*. These, I would remark, are precisely the relations which exist both on the northern and southern flanks of the Alps. M. Delbos further indicates, that his second stage in ascending order, which had been also confounded with the chalk, is a limestone characterized by *Schizaster rimosa*, *Hemiaster complanatus*, *Nummulina millecaput* (*N. gigas*, Catullo), *Serpula spirulæa*, most of which fossils occur in the shelly eocene of the Vicentine. Lastly, he points out, that although, even in his third or uppermost band, the *Ostrea lateralis* and the *O. gigantea* of the lower beds are repeated, they are there associated with a profusion of tertiary species. This band is the great receptacle of nummulites throughout the neighbourhood of Bayonne, the Corbières, &c., which nummulites (I may remark) are all or nearly all of the same species as in the Alps.

The facts developed by M. Leymerie are in my opinion essentially the same as those described by M. Delbos; for whilst he shows that the "terrain à nummulites" is connected with the chalk by help of certain fossils, still the great masses with nummulites are clearly superposed. But then this author has a theory to account for his "terrain épicrétacé." Seeing that these supposed secondary rocks of the south differ so much from those of the north of Europe, he explains this in his last memoir† by supposing that they were deposited in separate and distinct seas; so that certain animals may have continued to live on in the one, which had ceased to exist in the other basin. In this way he is inclined to think, that the nummulitic rocks of the south may represent at the same time the upper part of the cretaceous and the lower part of the tertiary system of the north.

To this I would reply by positive data. It has been shown that in this southern zone, and notably throughout the Alps, the very beds of transition or union are positively underlaid by the true equivalent of the white chalk and a full complement of the cretaceous system. Again, strata which M. Leymerie considers cretaceous, merely from the presence of the *Ostrea lateralis* and the *Terebratula tenuistriata*, are in my estimate the intermediate or transition beds only; and as the last-mentioned of these fossils is said to be undistinguishable from the *T. caput serpentis*, a species which mounts high into the tertiary deposits, nothing is gained by such an argument, particularly when the most secondary or cretaceous of the two species, the *Ostrea lateralis*, is stated to be associated with several well-known tertiary species.

In pointing out very clearly that the nummulitic rocks of Les Corbières are all posterior to the chalk, M. Talavignes‡ has endeavoured to divide the formation into what he calls two systems on account of their unconformity; but as no author has recognised a general break

* Bull. Soc. Géol. Fr. vol. iv. pp. 557, 713.

† Mém. de l'Académie de Toulouse.

‡ Bull. Soc. Géol. Fr. vol. iv. p. 1127.

even in the Pyrenees, I am disposed to consider this a local phænomenon, similar to that described by M. Favre in a portion of the Savoy Alps. It is needless, however, here to speak of lines of dislocation or transgressive deposits which I have disposed of elsewhere, as we are now merely dwelling on palæontological data and regular order of superposition; and the result of the researches of M. Talavignes is, that, with the exception of one Gryphæa, all the fossils of his two systems of nummulitic rocks are of tertiary forms.

An argument used by M. Dufrénoy to sustain the opinion of M. de Beaumont and himself, that the nummulite rocks formed the uppermost stage of the great cretaceous system of the south, has, it seems to me, fallen to the ground*. That author had indicated that the highly inclined nummulite strata of St. Justin in the Landes were surmounted by horizontal beds of calcaire grossier. On a scrutiny of this point, however, MM. Raulin and Delbos have proved, that the supposed calcaire grossier is a true Bordeaux miocene, and therefore we have there simply such a hiatus in the tertiary series as occurs in many parts of the Alps and Italy. M. Raulin has, indeed, gone further, and has proved, through the species of echinoderms†, that in the same region (Dax) there *is* a true equivalent of the white chalk, and that the overlying nummulitic rocks are loaded with eocene species. He insists, that whenever the nummulite group occurs, there is no other representative of the eocene. Hence M. Raulin believes that the great upheaval of the Pyrenees took place after the eocene epoch; and this is just what has occurred in the Alps. M. Rouant has, indeed, described a "terrain eocene" in the environs of Pau, which is the very same as the nummulitic group elsewhere, and being in an intermediate position, it is most satisfactory to know that it contains thirty-four fossil species of the Paris basin and five of the Vicentine.

Now, whatever these deposits in the south of France may be called, they are unquestionably of synchronous date with the nummulitic group of the Alps; for nearly every one of the same species of nummulites and orbitolites, besides many echinoderms and shells, occur in both regions in strata occupying the same place in the geological scale‡.

* Bull. Soc. Géol. Fr. vol. iv. p. 561. † Ibid. vol. v. p. 114.

‡ See p. 195, and the note on M. d'Archiac's identification of the species I brought from the Alps with those of the south of France. That able author has written to me, that he sees no zoological reason why that which he has termed the Asiatico-Mediterranean nummulitic group, extending, as he says, from the Asturias to the banks of the Brahmapootra, may not be the true type of the lower tertiary formation, whilst that which we have hitherto regarded as such (Paris, London, &c.) may have been due to local causes, and circumscribed to some ancient gulf of north-western Europe. What he still requires, before he modifies the opinions he has already expressed, are, clear proofs of geological and stratigraphical relations, and he hopes to find this point sustained in my memoir. Whilst speaking of the zoological characters of the nummulitic group, I am also happy to say, that a number of its fossils, forming part of a large collection in the Woodwardian Museum of Cambridge, procured from Count Münster, and ticketed by that naturalist from various Alpine localities cited in this memoir, have all been classed as eocene tertiary by Mr. F. M'Coy (the assistant of Professor Sedgwick) after a careful comparison of them with types of that age from other tracts.

To the very decisive opinion of M. Ewald of Berlin respecting the true tertiary character of the fossils of the Vicentine recorded in this memoir, I may add, that in a paper read before the Geological Section at Venice, he demonstrated that certain multilocular bodies in the hippurite limestone of Berre, near Marseilles, though resembling nummulites, were, in fact, quite distinct from them, both in structure and in the absence of the lenticular form. Abandoning his old opinions, like myself, M. Boué admits that as a whole the nummulites must be ranged in the eocene group, and he now the better understands why in certain parts of Turkey the miocene and younger tertiary at once succeed to nummulite rocks. It has indeed been stated by M. Constant Prevost, that nummulites occur with hippurites in the limestones of Cape Passaro in Sicily. That nummulite limestones immediately cover hippurite limestone in Italy, is a fact on which I have dilated; but whether the relations be the same in Sicily I cannot of course decide, not having been able to visit the spot. M. Coquand, whilst classing with the cretaceous rocks the nummulitic limestone and macigno of Morocco, shows at the same time, that the latter everywhere surmount the hippurite limestone; and this statement leads me to believe, that the general succession is the same in Africa as in Italy and in the Alps.

In casting our eyes eastward to the grand region of Northern Russia, we see how the deposits above the chalk preserve the type of our Northern Europe, and how in following them to the Carpathians and the Crimæa, they are found to assume the southern type. The sections of the nummulitic rocks of the south coast of the Crimæa, whether by M. Dubois or by M. de Verneuil, completely establish the fact, that the great mass of nummulitic limestone, with its *Ostrea gigantea* and other eocene fossils, is clearly superposed to the chalk. M. Dubois thinks, indeed, that one species of Nummulina there descends into the rock with true chalk fossils. But even if this be so, and that a true nummulite should also coexist with the uppermost hippurite rock of Cape Passaro in Sicily, it will only prove that the genus was called into existence a little earlier in those latitudes and longitudes than in the Alps and Apennines, whilst at the same time it would offer an additional proof of that very transition between the rocks called secondary and tertiary on which I have dwelt. However this may be, the facts remain the same, in relation to the *great masses of nummulites* that characterize the eocene of Southern Europe, which I have described. These, I repeat, are invariably supracretaceous; the nummulites being associated with a profusion of other animal remains of true tertiary character *.

* The superposition of true nummulites to the cretaceous rocks of the Asturias is announced to me by M. de Verneuil whilst these pages are passing through the press. The limestones and sandstones of that province which are charged with hippurites and radiolites, contain also abundance of *orbitolites*. The latter (which have been mistaken for nummulites) are fairly intercalated in the cretaceous system, and are surmounted by a yellowish limestone with spatangi, which may be the equivalent of the white chalk. This cretaceous group is distinctly overlaid by limestone abounding in true nummulites, which dips under sandstone and sands. This nummulitic band contains *Ostrea gigantea*, *Conoclypus conoideus*,

The nummulitic rocks which occupy large spaces in Egypt are all unquestionably of this same eocene age, as proved by their fauna. In a collection of fossils recently sent to the Royal Museum at Turin, M. Bellardi and myself recognized at a glance the eocene group of the Vicentine*.

Besides the *Nummulina millecaput* and *N. placentula*, well known in the Alps, these Egyptian rocks contain the *Bulla Fortisii*, Al. Brongn., *Turritella vittata*, Lamk., *T. imbricataria*, *Rostellaria fissurella* and *Nerita conoidea*, forms which are known in the Paris basin, in the Vicentine and at Nice. All the other Egyptian fossils, including Crustacea and Echinoderms, if not identical, are analogous to those of the supracretaceous group of the Alps and Italy. The same types of Pecten, of small, spinose Spondyli and Cardiaceæ, with Cassis and many univalves, complete the group.

Following this grand nummulitic formation from Egypt and Asia Minor† across Persia by Bagdad to the banks of the Indus, we long ago knew, from the communications of Capt. Grant, how in Cutch it is copiously loaded with fossils, which from the drawings and descriptions of Mr. James Sowerby‡ have all a tertiary aspect and relations. Subsequently the labours of Capt. Vicary, as recorded in our Proceedings§, have greatly added to our acquaintance with the range of these nummulitic rocks, which, in the form of limestones and sandstones, compose the great mass of the highly inclined strata of the mountain ranges of Hala and Solyman that separate Scinde from Persia, and extending from south to north, form the passes leading to Cabul. From collections recently sent home to me by Capt. Vicary it now appears, that some members of the same nummulitic group wrap round also from west to east in the Sub-Himalayan tracts in which Sabathoo is situated; and are said to reoccur, even in the kingdom of Assam. No geologist can view the fossils of this vast Eastern region (including nearly all the Punjaub, and even a large portion of Affghanistan) without being convinced that they belong to the same member of the series as the eocene of the Alps and Italy; for, with the same absence of ammonites, belemnites, hamites, or any cretaceous types, they exhibit six or seven species of num-

Serpula spirulæa, and other well-known eocene forms. The same order seems to prevail throughout Spain, even into the province of Malaga, and everywhere the nummulitic eocene, as in the Alps, has undergone the same flexures as the cretaceous rocks, whilst *the fossils* of the two formations are quite distinct.—June 1, 1849.

* Not more than half of this collection had been critically examined and compared when I left Turin in June 1848. I may here add, that a reference to Russegger's sections and description of the Mokattan Hills, near Cairo, would also lead inevitably to the inference, that the nummulitic rocks of Egypt are of eocene age (see Russegger, Reise in Europa, Asien und Afrika: Stuttgart; with fol. atlas). In short, this work affords evidence of the existence of true cretaceous rocks, followed by both eocene and younger tertiary deposits. Still M. Russegger, like most of his contemporaries, *classes the nummulite rocks with the chalk*.

† See Hamilton's Asia Minor, vol. i. pp. 405, 410, 500. M. Tchihatcheff will extend our knowledge on this point when he publishes the results of his recent travels.

‡ Trans. Geol. Soc. Lond. vol. v. Second Ser. p. 289 and plates.

§ Journal Geol. Soc. Lond. vol. iii. p. 331.

mulites*, four of which, the *Nummulina millecaput* or *polygyratus*, *N. planospira* or *assilinoides*, the *N. crassa* (Boubée), and the *N. Biaritzana* (D'Arch.), are identical with widely-spread and well-known forms of the South of Europe. Then again the same groups of radiata, conchifera and mollusca occur as in the nummulitic eocene of Europe. Some of the fossil shells of Scinde are, indeed, scarcely to be distinguished from the species of the Vicentine, particularly the so-called *Nerita conoidea* (Lamk.), or the *Neritina grandis* (Sow.), as well as one or two forms of Natica, the *Trochus agglutinans*, &c., whilst they have all a tertiary aspect, and if not identifiable with, are closely related to, our South European eocene forms†.

In comparing rocks of this epoch from distant parts of the globe, the amount of coincidence in their zoological contents is very remarkable, and in tracing their greater or less assimilation to our European types, we find, as might be supposed, that such is in a great measure dependent on the occurrence of similar or dissimilar conditions of deposit. Thus, in the Vicentine on the south flank of the Alps, where white limestones and marls abound, there are many more species common to that tract and the basin of Paris, than on the northern flank of the chain, where the deposits are more sandy and earthy, though their distance from the Parisian types is much smaller. Again, with the recurrence of strong resemblance to the lithological character of the Paris basin in the nummulitic rocks of Egypt and Scinde, we meet with a persistence of many identical or analogous forms, even at those vast distances. In the eyes of the geologist and palæontologist, therefore, the eocene type of Southern Europe extends through the heart of Asia, the differences in the fauna being simply characteristic of formations accumulated under varying conditions at the same time in distant seas. The surprise, indeed, is, that through the presence of certain species of nummulites, corals, echinoderms and shells, there should be so striking a resemblance in these widely separated deposits of so young an age as the eocene.

When we take the map of the world in hand, and view the enormous range of this nummulitic formation at intervals, *through twenty-*

* The researches of Capt. Vicary were undertaken by order of Lieut.-General Sir Charles Napier, after his brilliant conquest of Scinde. M. Leopold von Buch long ago recognized, in a letter to myself (see also Bull. Soc. Géol. Fr. vol. iv. p. 542), the identity of the nummulitic formation of Southern Europe with that which ranges from the Mediterranean and Egypt across Persia by Bagdad into Hindostan, and I much regret to have mislaid his short but pregnant sketch.

† Mr. Morris first examined these fossils of Scinde at my request, and seeing the close analogy which they present to the nummulitic group of Europe, had prepared a list of them. I have since submitted them to M. d'Archiac at Paris, in the hope that he may describe them in detail for the Geological Society of London, and compare them with the nummulitic fauna of Southern France, which he has well studied. The species named with the aid of Mr. Morris, in the Table at the end as having a wide range, result in part from these examinations, and also from a comparison of the corals by M. Jules Haime (the associate of M. Milne Edwards), who has stated that four species of that class derived from Scinde, are identical with forms published from Nice. I may also add, that I saw in the Royal Museum at Turin, a Cyclorite from the mountains between Scinde and Cabul, which M. Bellardi identifies with the *Cyclolites Borsoni* (Michelin) of Nice.

five degrees of latitude and *near one hundred degrees of longitude,* its northernmost ridge on the north flank of the Carpathians being clearly identifiable with its southernmost known limb in Cutch, and its western masses in Spain and Morocco being similar to those of the Brahmapootra, we at once see the vast importance which attaches to a right understanding of its true place in the geological series. And this assimilation of distant deposits is effected, it will be remembered, in spite of great local diversities of lithological and mineral character. The black subcrystalline schists and limestones of the summits of the Vallaisan and Savoyard Alps, with their Cerithia and Melaniæ, and the black fish-slates of Glarus; the hard, calcareous, green sandstones of the Alps of Bern, of the four cantons, and of Bavaria, are all proved by their fossils and order of superposition to have been formed during the same geological period as the white limestones, marls and sandstones of Monte Bolca and the Vicentine, and by zoological inference, at the same time as similar rocks in Egypt and Hindostan. Nay more, we see in the Alps enormous thicknesses of overlying "flysch" and "macigno," which having often the aspect of the oldest secondary or even of transition rocks, are not of higher antiquity than our unconsolidated London clay and Bagshot sands!

In coming to my present opinion I regret to be compelled to dissent from my eminent friend M. Élie de Beaumont; for even in the last modification of his opinions, he views the "terrain à nummulites" as a member of the cretaceous rocks. In one essential point indeed, when he states that complete researches will probably make known passages or transitions between all conterminous formations, he gives the great value of his sanction to opinions I have long held and published*. I rejoice that he pointedly adverts to the error of those who believe in *general dislocations,* or revolutions which have neatly separated one great group of rocks and their imbedded animals from another; and that stating how all disruptions are local in reference to the surface of the globe, he admits with me, that even in two formations unconformable to each other, some of the same organic remains have been found to exist. Apparently, however, not sufficiently acquainted with the presence in the Alps of a full representative of the *chalk,* and believing that the nummulitic series there rests upon strata of the age of the greensand, he supposes that the nummulitic group and flysch of that chain may answer to the upper part of the cretaceous system, and may also fill up the interval so frequently observable in Northern Europe, between the surface of the chalk and the plastic clay. But he must forgive me when I state my belief, that this view cannot now stand in the face of the clearly-ascertained succession which has been pointed out. If it were valid, then the nummulitic rocks and flysch or "terrain épicrétacé" would surely somewhere be overlaid by a zoological representative of the calcaire grossier; whereas in every country where it is known, the nummulitic and flysch group is surmounted, for the most part unconformably, by deposits with miocene or pliocene shells. Even if

* See Silurian System and Russia in Europe, *passim.*

they were void of fossils, the enormous accumulations of finely-laminated beds, which overlie the true equivalents of the chalk, and are followed by the deposits which hitherto have alone been viewed as younger tertiary, must represent so long a period, that as physical monuments only they are in my mind's eye, full and complete equivalents *in time* of the eocene of geologists*.

And now a word upon the reform which the adoption of this view must introduce into geological maps. The truth is, that in previous classifications of the rocks of Southern Europe the eocene formation has been almost omitted, chiefly because it there usually forms the upper portion of a continuous and unbroken series of strata, of which the neocomian limestone or lowest member of the cretaceous system is the base. In some tracts it will doubtless be difficult, except the scale of the map be large, to indicate the separation of the eocene from such cretaceous rocks; but on the other hand, it will be as easy as it is necessary to mark this formation by a distinct colour over enormous spaces, separating it from the cretaceous on the one hand and from the younger tertiary deposits on the other. Even in the most general maps I conceive that this distinction may be effected. No geological division can, indeed, be more essential than that which distinguishes lower tertiary rock-masses from those of upper secondary age; inasmuch as, with the exception of certain beds of junction, the two groups have no organic remains in common, and afford the clearest proofs of having been formed at different periods of time, and when the submarine fauna underwent a total change.

Lastly, let me say, that without taking a comprehensive view of the whole question, and alluding to the works of my contemporaries, I should not have made apparent the value of the establishment of a clear order of secondary and tertiary succession in the Alps, Carpathians and Italy. In respect to my leading object, I repeat, that wherever true and full representatives of the different members of the cretaceous system occur, from the neocomian or equivalent of the

* In a letter recently received from M. Alcide d'Orbigny, he thus expresses himself: "For three years I have made the most extensive researches upon Nummulites; and in comparing all the stratigraphical and palæontological results, it is impossible not to recognize therein two distinct epochs, as represented by strata, superposed the one to the other, and having each its proper fauna. One of these epochs, which I have recognized in the French Alps, the Pyrenees and the Gironde, corresponds to the plastic clay of Paris and London, and which, belonging to the lower sands of Soissons, I have named 'Étage Suessonien'; the other, equally common in the Alps and the basins of the Gironde, and which includes the 'calcaire grossier' of Paris up to the gypsum of Montmartre and the London clay, &c., I designate 'Étage Parisien.' These divisions, based upon a considerable number of facts, are detailed in the work I am now printing, and the entire composition of their characteristic faunas is given in my 'Prodromus of Universal Palæontology.' The habit I have acquired of determining these fossils makes me regret that I cannot go to inspect your collections in London; but the portions of them I have seen in the hands of our friend M. de Verneuil have led me to recognize at once what I was already acquainted with in the Pyrenees and the French Alps. Again, the fossils I have examined in the collection of M. Tchihatcheff (recently brought from Asia Minor) confirm me in my opinion, and would lead me to extend the limits of these tertiary stages, as you have suggested, even to Hindostan."

lower greensand, upwards through the gault and upper greensand into the white chalk inclusive,—there also all the species of the genus *Nummulina* lie invariably above such strata; and further, that with the exception of one or two forms of Gryphæa and Terebratula (conchifers peculiarly tenacious of life, and which generally occur in the beds of transition above the chalk, and never rise above the lower beds of the nummulitic group), all the fossils associated with the nummulites are of eocene type. I am glad that these conclusions, derived from geological researches and absolute sections, are in harmony with the results obtained by the most eminent naturalists from their study of organic remains. Brongniart, Deshayes and D'Orbigny have long maintained that the nummulites of France are truly of tertiary age. Agassiz groups them as rather pertaining to a peculiar or lower tertiary. In his recent valuable tabular view of all known fossils (to which I specially invite attention), Professor Bronn of Heidelberg places the nummulitic group as the natural base of all the tertiary deposits. This concordance of physical geology with palæontology has indeed been everywhere established where patient researches have been carried out.

In conclusion, it is unnecessary that I should revert to all the deductions I have attempted to draw concerning the operations of metamorphism, contortion, and fracture by which the strata of the Alps and Apennines have been so powerfully affected; and I will now simply recapitulate the chief points which I have grouped together, in presenting to my countrymen a view of the normal order of the formations, as well as of the derangements they have undergone, in the Alps, Carpathians and Apennines.

1. That whilst evidences of Silurian, Devonian and carboniferous rocks exist in the Eastern Alps, the palæozoic group of Southern Europe nowhere exhibits traces of the Permian system of Northern Europe.

2. That these palæozoic rocks are succeeded in the Eastern Alps, and notably in the South Tyrol, by the "Trias," as characterized by known muschelkalk fossils and also by many species peculiar to the Alpine zone of this system; whilst none of these fossils have yet been recognized in the Western Alps.

3. That the Jurassic system of the Alps and Apennines is made up of two distinct calcareous formations; the inferior representing the lias and lower oolites, the superior the Oxfordian group, so largely developed throughout Russia, though in a very different mineral condition.

4. That the cretaceous system of Southern Europe is composed of hard subcrystalline Neocomian limestones (the equivalents in great part of the English lower greensand), of a band replete with fossils of the gault and upper greensand, and of red, grey and white limestones with Inocerami representing the chalk.

5. That where the sequence is full and unbroken, the cretaceous rocks of the Alps and Apennines graduate conformably and insensibly upwards by mineral and zoological passages into the nummulitic zone,

in which and in its great intercalated and overlying masses of flysch or upper "macigno" the secondary types have vanished, and an eocene tertiary fauna appears.

6. That by the presence of numerous fossils, and notably by its nummulites and echinoderms, this eocene group is known to extend from the Mediterranean through Egypt, Asia Minor and Persia to Hindostan, and there to occupy large regions forming the western and northern limits of British India.

7. That the names of Carpathian sandstone and Vienna sandstone, as well as of flysch and macigno, have been applied to rocks which are both of secondary and tertiary age; but that in the Carpathians, as in the Alps, those portions of them containing nummulites with certain overlying strata represent the eocene tertiary.

8. That the cretaceous and nummulitic eocene formations of the Alps having been successively deposited under the sea, have since undergone the same common flexures and fractures, by which the younger strata have been frequently folded under those of older date.

9. That the only general feature of independence in the formations of the Northern Alps, is that which is exhibited in the grand rupture and hiatus between the pre-existing nummulitic eocene with flysch and the subsequently-formed molasse and nagelflue.

10. That as the marine contents of the Swiss molasse, whether called younger miocene or older pliocene, exhibit a large proportion of living species of marine shells, whilst the associated and *overlying* strata of terrestrial origin, often called molasse, are loaded with forms all of which are extinct, the same terms cannot be applied as equivalents to define the tertiary strata which were formed contemporaneously under the sea and upon the land (see p. 237).

11. That although on the southern flank of the Venetian Alps the nummulitic eocene group is followed by younger tertiary deposits, which, also elevated at high angles, have a direction parallel to the older chain, it is believed that such external lower parallel (Bassano, Asolo) was produced after that chief elevation which raised the secondary and eocene rocks together, and has in many places left the latter upon the summits of the Alps.

12. That notwithstanding local dislocations, Northern Italy further exhibits conformable passages from what may be the uppermost eocene or lowest miocene high up into subapennine strata, in which most of the shells are undistinguishable from those now living.

13. That since the emersion of all the pliocene and youngest marine deposits and their addition to the pre-existing lands, the oscillations which the coasts of Italy have undergone, particularly during the historic æra, are symptoms of the remains only of that subterranean energy which was exerted with much greater intensity during former periods in the Alps, Carpathians and Apennines.

Species of the Nummulitic Eocene group having a wide geographical range.

Fossils.	Localities.
Nummulina millecaput, *Boubée*=*N. polygyratus*, Desh.	Alps. Pyrenees. Crimea. Egypt. Vicentine. Scinde.
—— planospira, *Boubée*=*N. assilinoides*, Rüt.	South of France. Pyrenees. Alps. Apennines. Carpathians. Mt. Gargano (Naples).
—— Biaritzana, *d'Arch.*=*N. atacica*, Leym.=*N. acuta*, Sow.=*N. regularis*, Rüt.	Alps. Biaritz. Vicentine. Cutch and Scinde.
—— rotularis, *Desh.*=*N. globulus*, Leym.=(*N. lævigata*, Pusch, t. 12. f. 16 *a*)?	South of France. Pyrenees. Alps. Crimea. Carpathians?
—— placentula, *Desh.* = *N. intermedia*, d'Arch.	South of France. Alps. Crimea. Egypt. Scinde?
—— globosa, *Rüt. & d'Arch.*=*N. obtusa*, Joly & Leym. (var. of *Biaritzana*).	Alps. South of France.
—— lævigata, *Lamk.*	London. Paris. Belgium. Lower Pyrenees. Vicentine. Mt. Gargano (Naples).
—— granulosa, *d'Arch*..........	Dax. Pyrenees. Mt. Gargano (Naples). Asia Minor.
—— crassa, *Boubée* = *N. obtusa*, Sow.	Alps. Pyrenees. Cutch.
Orbitolites submedia, *d'Arch.*=*O. Prattii*, Michelin.	South of France. Pyrenees. (Matsee.) Alps.
—— discus, *Rüt.*	South of France. Alps. Scinde.
—— patellaris, *Brunner*	Alps.
—— stellaris, *Brunner*=*Calcarina stellata*, d'Arch.	Swiss Alps. Vicentine. Nice. South of France.

Zoophyta.

Trochocyathus bilobatus, *M. Edwards and J. Haime, Ann. Scien. Nat.* 3 ser. vol. ix. p. 331.	Nice. Scinde.
—— multisinuosus, *M. Edwards and J. Haime, ibid.* p. 336.	Nice. Scinde.
—— near to *T. cyclolitoides, M. Edwards and J. Haime, ibid.*	Scinde.
Trochosmilia corniculum, *M. Edwards and J. Haime, ibid.* p. 240.	Nice. Scinde.
Stylocænia emarciata, *M. Edwards and J. Haime, ibid.*	Paris. Scinde.
Ceratotrochus near to *C. exaratus, M. Edwards and J. Haime, ibid.*	Scinde.
Cyclolites Borsoni, *Michelin*	Rivalta (Bormida). Nice. Scinde.
Astræa radiata, *Lamk.*	Paris. Vicentine. Rivalta (Bormida).
Meandrina profunda, *Michelin*	Vicentine. Rivalta (Bormida).

Obs.—The greater number of the corals of the Vicentine have not yet been compared with those of other localities.

Fossils.	Localities.
RADIARIA.	
Pygorhynchus Cuvieri, *Münst.* sp. .	Paris. N. Alps.
—— subcylindricus, *Ag.*	Trent. Pyrenees.
Conoclypus conoideus, *Lamk.* sp...	N. Alps. (S. Alps.) Pyrenees. Asturias. Nice. Vicentine. Crimea. Egypt.
Echinocyamus profundus, *Ag.*	Trent (S. Tyrol). Swiss Alps.
Echinolampas politus, *Ag.*........	N. Alps. South of France.
—— subsimilis, *d'Arch.*	Pyrenees. Trent (S. Tyrol). Cutch.

Obs.—The number and variety of the species of Echinoderms, chiefly elongated, which are found in the nummulitic group in the Alps, Pyrenees and India, amounting to upwards of 100 species, eminently characterize this formation; not one of them being known in the cretaceous rocks. The greatest number of species belong to the genera Echinolampas, Conoclypus, Pygorhynchus, Eupatagus, Hemiaster and Schizaster (see Agassiz).

Fossils.	Localities.
CRUSTACEA.	
Cancer Sonthofensis	Sonthofen, Bavarian Alps.

Obs.—Other species of Crustacea are also abundant in the Alps, Egypt, Scinde, &c.

Fossils.	Localities.
ANNELIDA.	
Serpula spirulæa, *Lamk.*	Paris. Swiss and Bavarian Alps. Vicentine. Asturias.
CONCHIFERA.	
Cytherea elegans, *Lamk.*..	London. Paris. Vicentine.
Venericardia acuticostata, *Lamk.*= *V. Lauræ*, Brong.= *Cardium semigranulatum*, Münst.	Paris. Vicentine.
—— minuta, *Leym.*	Pyrenees. Nice. Egypt.
Chama squamosa, *Sow.*..........	London. Bassano.
Pholadomya Puschii, *Goldf.*......	London? South of France. Westphalia. Nice. Vicentine. Scinde.
Crassatella sulcata, *Sow.*	London. Schio. Vicentine.
Pecten corneus, *Sow.*=*P. suborbicularis*, Münst.	London. Kressenberg. Swiss Alps.
—— plebeius, *Lamk.*............	Paris. Kressenberg. Swiss Alps.
—— scutularis, *Lamk.*	Paris. Kressenberg. Swiss Alps.
Ostrea gigantea, *Dubois*=*O. latissima*, Desh.	London. Paris. South of France. Nice. Vicentine. Alps. Pyrenees. Asturias. Crimea.
—— multicostata, *Desh.*	Paris. Pyrenees. Nice. Egypt.
Terebratula bisinuata, *Desh.*=*T. subalpina*, Münst.	Paris. Kressenberg.
Spondylus cisalpinus, *Brong.*	Nice. Sardagna near Trent. (S. Tyrol.) Vicentine and Bavarian Alps.
MOLLUSCA.	
Conus diversiformis, *Desh.*	Paris. Scinde.
—— stromboides (=*C. concinnus*, Sow.).	London. Bassano and Vicentine.

Fossils.	Localities.
Ovula tuberculosa, *Duclos*	Paris. Crimea. Scinde.
Voluta Cithara, *Lamk.*	Paris. Scinde.
—— harpula, *Lamk.*	Paris. Bassano.
Bulla Fortisii, *Brong*............	Vicentine. Egypt.
—— striatella, *Lamk.*	Paris. Vicentine.
Terebra Vulcani, *Brong.*	Vicentine. Scinde.
Cerithium giganteum, *Lamk.*?	London. Paris. Venetian Alps. Nice. Crimea. Scinde, &c.
—— hexagonum, *Lamk.*=*C. pentagonum*, Fortis = *C. Maraschini*, Brong.	Paris. Cotentin. Vicentine.
—— cornucopiæ, *Lamk.* = *C. armatum*, Münst.	Paris. Cotentin. Vicentine.
Rostellaria fissurella	Paris. Nice. Vicentine. Egypt.
Strombus Fortisii, *Brong*.........	Vicentine. Scinde.
Fusus longævus, *Lamk.*..........	London. Paris. Vicentine. Bassano.
—— intortus, *Lamk.*............	Paris. Bassano.
Neritina conoidea, *Lamk.*	Paris. Pyrenees. S. Tyrol. Vicentine. Egypt. Scinde.
Natica sigaretina, *Lamk.*	Paris. Nice. Vicentine. Scinde, &c.
Pleurotoma semicolon, *Sow.*......	London. Bassano. Possagno, &c.
—— undata, *Lamk.*	Paris. Bassano.
Melania costellata, *Lamk.*........	Paris. Swiss Alps. Vicentine.
—— lactea, *Lamk.* = *M. Stygii*, Brong.	Paris. Vicentine.
Turritella Archimedis, *Brong.*	Paris. Pyrenees. Egypt.
—— imbricataria, *Lamk.*	London. Paris. Swiss Alps. Vicentine. Egypt. Crimea. Scinde.
—— vittata, *Lamk.*	Paris. Nice. Vicentine. Egypt. Scinde.
Trochus monilifer, *Lamk.*........	Paris. Scinde.
—— agglutinans, *Lamk.*	Paris. Vicentine. Scinde.
Nautilus ziczac, *Sow.*............	London. Kressenberg. Matsee.

Obs.—Among the fossils recently sent to me by Capt. Vicary from Subathoo in Hindostan, are fragments of the lower jaw and teeth of a small gavial, of which Professor Owen says: "It seems to have rather rounder teeth than the modern species in India, and in this respect to resemble our old British eocene gavial of Bracklesham." None of the other forms from this Sub-Himalayan tract (according to Professor E. Forbes, to whom I referred them,) indicate the presence of rocks more ancient than the nummulitic eocene.

Under the term Scinde, &c. the reader may comprehend Cabul, the Punjaub, the valley of Cashmir, and the Sub-Himalayan range to the kingdom of Assam. Mr. Vigne, who explored Cashmir, has shown me limestone charged with nummulites from thence.

Postscript.—In addition to my own limited observations on the Trias of the Venetian and S. Tyrolese Alps (p. 165), I intended to have referred my readers to the illustration of the rocks and fossils of that age contained in the work of Professor Catullo, "Prodromo di Geognosia paleozoica delle Alpi Venete. Modena, 1847." Besides the common muschelkalk species cited in the preceding pages, Professor Catullo figures and describes several new species, and also the interesting triassic plant *Voltzia brevifolia* (Brong.). He further enumerates many fossils of the jurassic and cretaceous groups of that region, and figures their Cephalopoda. I cannot pretend to decide authoritatively a point on which this author insists—that certain species are common to the Upper Jura and Neocomian; but whilst I should be very sorry to do injustice to so experienced a naturalist as Professor Catullo, I must repeat, that wherever I have examined a tract in which there was a clear geological succession, there also the accompanying zoological distinctions indicated by M. de Zigno seemed to me to be equally clear. In cases of this nature everything depends upon correct definitions of the relations and order of the strata. Professor Catullo also describes five species of nummulites from the tertiary rocks of the Vicentine, but I must leave others to determine how far these forms have been named by previous authors. In another work ("Cenni sopra il terreno di sedimento superiore Venezia. 1847."), Professor Catullo figures a number of tertiary corals.

I have just received a new geological map of the environs of Vienna by M. Johann Czjzek, in which the author represents the "Wiener Sandstein" as older than the Alpine (jurassic?) limestone! I have not sufficiently re-examined that tract to be able to controvert this inference, but I firmly hold to the facts stated in the preceding memoir; and as the Bavarian "flysch" is unquestionably, like that of Switzerland, supracretaceous, it is for the Austrian geologists to show that their "Wiener Sandstein" is neither a prolongation of the same deposit, nor even an arenaceous development of any portion of the cretaceous system.

N.W.

Sections

across

THE HOHER-SENTIS

drawn by

M. Arnold Escher von der Linth

to illustrate a memoir by

Sir Roderick I. Murchison.

Section 2.

Section 3.

Section 4.

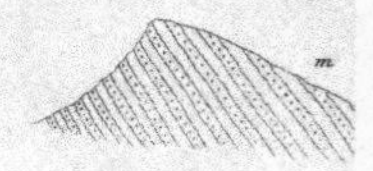

Section 5.

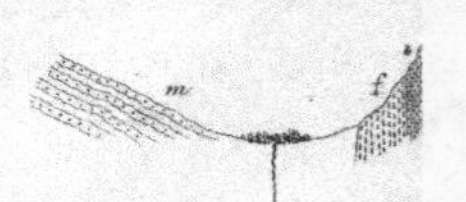

Section 7.

Section

Molasse & Nagelflue

Eocene (Murchison)

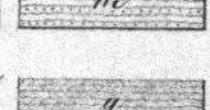

Flysch

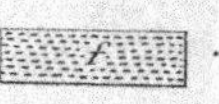

Nummulite limestone

(transition)

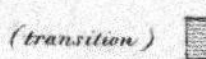

Gryphite beds

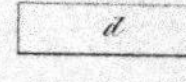

Inoceramus limestone (Chalk)

Cretaceous System

Upper Green Sand & Gault

Upper Neocomian

Lower Neocomian

(The transition beds (e) have been inserted at the suggestion of Sir R. Murchison.)

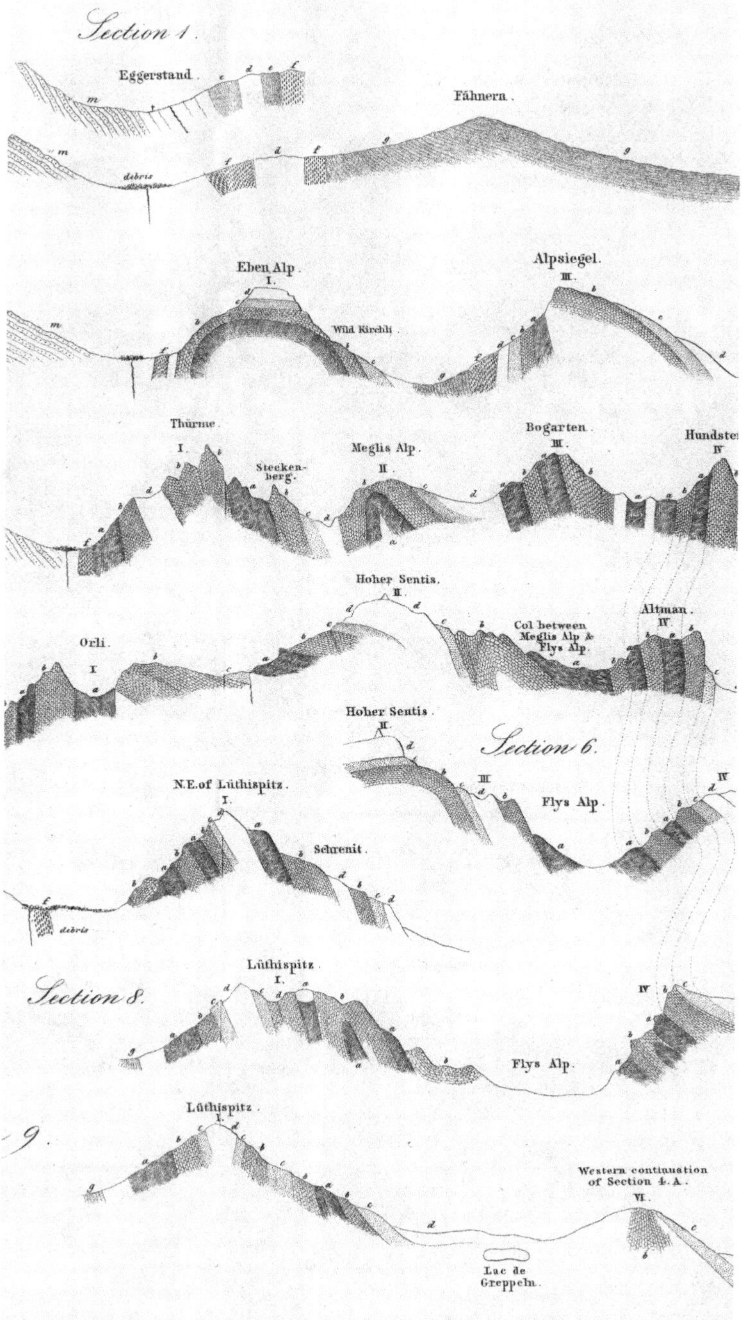

Section 1.

Eggerstand.

Fähnern.

debris

Eben Alp.
I.

Wild Kirchli.

Alpsiegel.
III.

Thürme.
I.

Stecken-
berg.

Meglis Alp.
II.

Bogarten.
III.

Hundste
IV

Hoher Sentis.
II.

Orli.
I

Col between
Meglis Alp &
Flys Alp.

Altman.
IV.

Hoher Sentis.
II.

Section 6.

III

IV

N.E.of Lüthispitz.
I.

Flys Alp.

Schrenit.

debris

Lüthispitz.
I.

Section 8.

IV

Flys Alp.

Lüthispitz.
I.

9.

Western continuation
of Section 4. A.
VI.

Lac de
Greppeln.

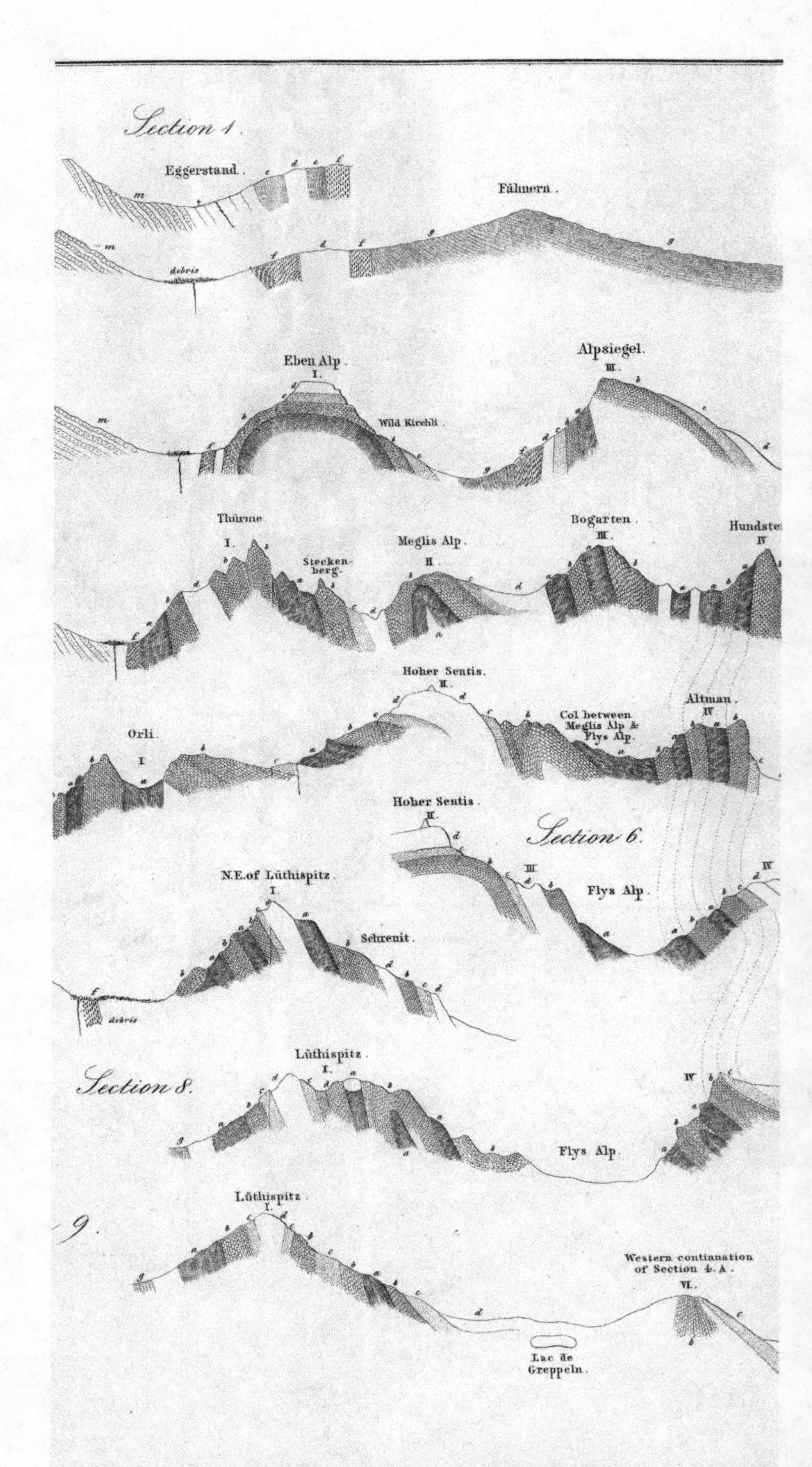
Section 1.
Eggerstand.
Fähnern.
debris
Eben Alp.
I.
Wild Kirchli.
Alpsiegel.
III.
Thürme
I.
Stecken-berg.
Meglis Alp.
II.
Bogarten.
III.
IV
Hoher Sentis.
II.
Orli.
I
Col between Meglis Alp & Flys Alp.
Altman.
IV.
Hoher Sentis.
II.
Section 6.
III
IV
N.E. of Lüthispitz.
I.
Schrenit.
Flys Alp.
debris
Lüthispitz.
I.
Section 8.
IV
Flys Alp.
Lüthispitz.
I.
9.
Western continuation of Section 4. A.
VI.
Lac de Greppeln.

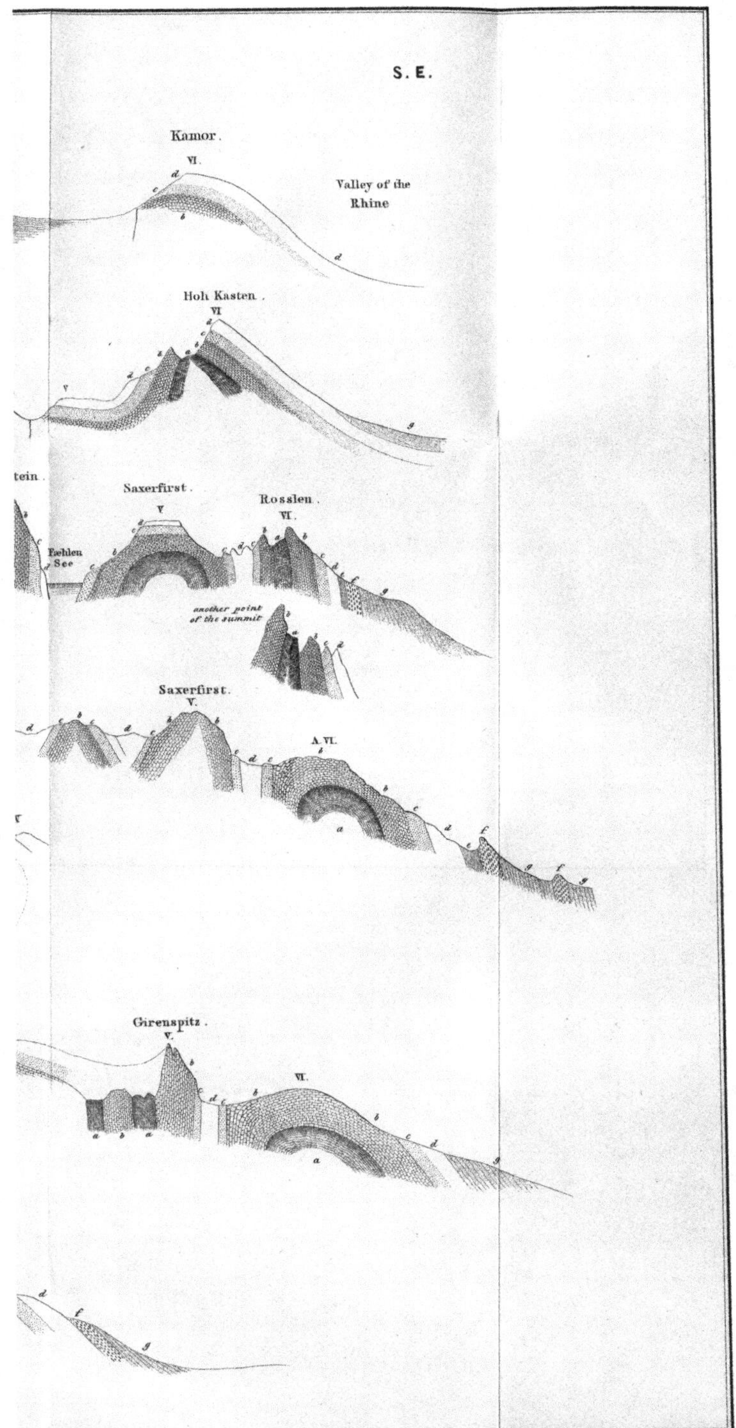

Reeve, Benham & Reeve, Lithographers.

[From the LONDON, EDINBURGH, and DUBLIN PHILOSOPHICAL MAGAZINE AND JOURNAL OF SCIENCE for July 1849.]

On the Distribution of the Superficial Detritus of the Alps, as compared with that of Northern Europe. By Sir RODERICK IMPEY MURCHISON, *F.R.S., V.P.G.S. &c.; Mem. Imp. Ac. Sciences of St. Petersburgh, Corresp. Member of the Academies of Paris, Berlin, Turin, &c.*

REFERRING to his previous memoir upon the whole structure of the Alps and the changes which those mountains underwent, the author calls attention to the fact, that whilst during the formation of the molasse and nagelflue a warm climate prevailed, so after the upheaval of these rocks an entire change took place, as proved by the uplifted edges of these tertiary accumulations being surmounted by vast masses of horizontally-stratified alluvia, the forms of whose materials testify that they were deposited under water. The warm period, in short, had passed away and the pine had replaced the palm upon the adjacent lands, before a glacier was formed in the Alps or a single erratic block was translated.

Though awarding great praise to the labours of Venetz, Charpentier and Agassiz, which have shed much light on glaciers, and particularly to the work of Forbes for so clearly expounding the laws which regulate the movement of these bodies, Sir Roderick conceives, that the physical phænomena of the Alps and Jura compel the geologist to restrict the former extension of the Alpine glaciers within infinitely less bounds than have been assigned to them by those authors. True old glacier moraines may, he thinks, be always distinguished, on the one hand, from the ancient alluvia, and on the other from tumultuous accumulations of gravel, boulders and far transported erratic blocks, as well as from all other subsequent detritus resulting from various causes which have affected the surface. He first shows, from the remnants of the old water-worn alluvia which rise to considerable heights on the sides of the valleys, that in the earliest period of the formation of the Alpine glaciers, water, whether salt, brackish or fresh, entered far into the recesses of these mountains, which were then at a considerably lower level, *i. e.* not less than 2500 or 3000 feet below their present altitude.

He next appeals to the existing evidences in the range of Mont Blanc to show, that as each glacier is formed in a *transverse* upper depression, and is separated from its neighbour by an intervening ridge, so by their movement such glaciers have always protruded their moraines across the adjacent longitudinal valleys into which they descended—and were never united to form one grand stream of ice. It is stated that there are no traces of lateral moraines on the sides of the main valleys at considerable heights above their present bottoms, whether on the flank of the great ridge from whence the glaciers issued or on the opposite side of each longitudinal valley, which must have been the case if a large mass of glacier ice had ever descended the general valley. On the contrary, examples of the transport of moraines and blocks *across* such *longitudinal* depressions are cited from the valley of Chamonix on the one flank and from the Allée Blanche and Val Ferret on the other flank of the chain of Mont Blanc. Another proof is seen in the ancient moraine of the Glacier Neuva, the uppermost of the valley of the Drance; and a still stronger case is the great chaotic pile of protogine blocks accumulated on the Plan y Bœuf, 5800 French feet above the sea, which have evidently been translated right across the present deep valley of the Drance, from the opposite lofty glacier of Salenon.

Having thus shown that none of the upper longitudinal and flanking valleys around Mont Blanc were ever filled with general ice-streams, the author has still less difficulty in demonstrating that all the great trunk or lower valleys of the Arve, the Doire, and the Rhone, offer no vestiges of what he calls a true moraine; all the detritus from great heights above their present bottoms exhibiting either water-worn pebbles or occasional large erratic blocks, more or less angular,—the latter being for the most part irregularly and sporadically dispersed. As Venetz and Charpentier have attached great importance to the original suggestion of an old peasant of the Upper Vallais, that a great former glacier alone could have carried the erratic blocks to the sides of the lower valley of the Rhone, so on the other hand the author relies on the practised eye of his intelligent Chamonix guide Auguste Balmat, who declares that he has never recognized the remains of "moraines" in that detritus of the larger valleys which has been theoretically referred to glacier action. In descending from the higher Alps into such trunk valleys, Sir Roderick found many examples of rocks rounded on the side which had been exposed to the passage of boulders and pebbles, with abrupt faces on the side removed from the agent of denudation, all of them reminding him forcibly of the storm and lee sides of the Swedish rocks over which similar water-worn materials have passed.

Seeing, then, that this coarse drift or water-worn detritus is distributed sometimes on the hard rocks and often on the summits of the remnants of the old valley alluvia, he believes that the whole of the phænomena can be explained by supposing that the Alps, Jura, and all the surrounding tracts have undergone great and unequal elevations since the period of the formation of the earliest glaciers—elevations which, dislodging vast portions of those bodies, floated away many huge blocks down straits then occupied by water, and hurled on vast turbid accumulations of boulders, sand and gravel. To these operations he attributes the purging of the Alpine valleys of the great mass of their ancient alluvia, and also the conversion of glacier moraines into shingle and boulders. He denies that the famous blocks of Monthey opposite Bex, can ever have been a portion of the left lateral moraine of a glacier which occupied the whole of the deep valley of the Rhine,—as Charpentier has endeavoured to show;

and he contends that if such had been the case they would have been associated with numberless smaller and larger fragments of all the rocks which form the sides of the valley through which such glaciers must have passed. They are, however, exclusively composed of the granite of Mont Blanc; and must therefore, he thinks, have been transported by ice rafts,—which, having been forced with great violence through the gorge of St. Maurice, served to produce many of the striæ which are there so visible on the surface of the limestone*. Fully admitting that the stones and sand of the moraines of modern glaciers scratch, groove, and polish rocks, Sir Roderick Murchison still adheres to the idea he has long entertained from surveys in Northern Europe, that other agents more or less subaqueous, including icebergs and great masses of drift, have produced precisely similar results. He cites examples in the Alps, where perfectly water-worn or rounded gravel being removed, the subjacent rocks are found to be striated in the directions in which such gravel has been moved; and he quotes a case in the gorge of the Tamina, above the Baths of Pfeffers, where this ancient striation, undistinguishable from that caused by existing glaciers, has, by a very recent slide of a heavy mass of gravel from the upper slope of the same rock, been crossed by fresh scorings and striæ, transverse to those of former date, from which the markings made in the preceding year only differ in being less deeply engraved. He also adverts to the choking up of some valleys, particularly of the Vorder Rhein below Dissentis, by the fracture, *in situ*, of mountains of limestone, which constitute masses of enormous thickness, made up of innumerable small fragments, all of which have been heaped together since the dispersion of the erratic blocks; and he further indicates the effects of certain great slides or subsidences within the historic æra.

In considering the distribution of the erratic detritus of the Rhone, Sir Roderick having denied that it can ever have been carried down the chief valley to the Lake of Geneva in a solid glacier, still more insists on the incredibility of such a vast body of ice having issued from that valley, as to have occupied all the low country of the cantons Vaud, Friburg, Berne and Soleure, and to have extended its erratics to the slopes of the Jura, over a region 100 miles in breadth from north-east to south-west as laid down in the map of Charpentier. He maintains that in the low and undulating region between the Alps and the Jura, the small debris derived from the former has everywhere been water-worn, and that there is in no place anything resembling a true moraine; and he therefore believes, that the great granitic blocks of Mont Blanc were translated to the Jura by ice-floats, when the intermediate country was under water. He further appeals to the water-worn condition of all the detritus of the high

* Mr. Charles Darwin, in a recent letter to the author, adheres to his old opinions derived from observations in America, and says, "I feel most entirely convinced that *floating ice* and *glaciers* produce effects so similar, that at present there is, in many cases, no means of distinguishing which formerly was the agent in scoring and polishing rocks. This difficulty of distinguishing the two actions struck me much in the *lower parts* of the Welsh valleys."

plateaux around Munich, 1600 and 1700 feet above the sea, to show that a subaqueous condition of things must be assumed when the great erratic blocks were carried to their present positions.

Prof. Guyot of Neufchatel has endeavoured to show, that the detritus of the rocks of the right and left sides of the upper valley of the Rhone have also maintained their original relative positions in the great extra Alpine depression, and that these relations are proofs, that nothing but a solid glacier could have arranged the blocks in such linear directions. But the author meets this objection by suggesting that there are notable examples to the contrary. He also refers to the great *trainées* of similar blocks which preserve linear directions in Sweden and the low countries south of the Baltic, to show that as this phænomenon was certainly there produced by powerful streams of water, so may the Alpine detritus have been arranged by similar agency. In alluding to the drainage of the Isère he further points to the admission of Prof. Guyot, that nearly all its erratic detritus, both large and small, is rounded and has undergone great attrition; and he quotes a number of cases in which such boulders and gravel, derived from the central ridges of Mont Blanc, have been transported *across* tracts now consisting of lofty ridges of limestone with very deep intervening valleys; and therefore he infers that the whole configuration of these lands has been since much changed, including the final excavations of the valleys and the translation of enormous masses of broken materials into the low countries of France.

In conclusion it is suggested, that the dispersion of the far-travelled Alpine blocks is a very ancient phænomenon in reference to the historic æra, and must have been coeval with the spread of the northern or Scandinavian erratics, which it has been demonstrated was accomplished chiefly by floating ice, at a time when large portions of the Continent and of the British Isles were under the sea. Viewing it therefore as a subaqueous phænomenon, Sir Roderick is of opinion that the transport of the Alpine blocks to the Jura falls strictly within the dominion of the geologist, who treats of bygone events, and cannot be exclusively reasoned upon by the meteorologist, who invokes a long series of years of sunless and moist summers to account for the production of gigantic glaciers upon land. This last hypothesis is at variance even with the physical phænomena in and around the Alps, whilst it is in entire antagonism to the much grander and clearly established distribution of erratics of the North during the glacial period. The effect in each case is commensurate with the cause. The Scandinavian chain, from whence the blocks of central Europe radiated, is of many times larger area than the Alps, and hence its blocks have spread over a much greater space. All the chief difficulties of the problem vanish when it is admitted, that enormous changes of the level of the land in relation to the waters have taken place since the distribution of large erratics; the great northern glacial continent having subsided, and the bottom of the sea further south having been elevated into dry land, whilst the Alps and Jura, formerly at lower levels, have been considerably and irregularly raised.

Note on the Geological Structure of the Asturias, particularly in reference to the Nummulitic Eocene, and the Carboniferous Palæozoic Rocks of that Province (extracted from a letter of M. E. DE VERNEUIL *addressed to* Sir RODERICK I. MURCHISON).

IN a tour which he is now making in Spain, M. Ed. de Verneuil has observed, that on the frontiers of the provinces of Asturias and Santander, the nummulitic formation overlies all the true cretaceous rocks, and that no form of the genus *Nummulina*, D'Orb., ever occurs in them; thus fortifying the generalization recently announced by Sir Roderick Murchison, deduced from a study of the Alps, Apennines, and Carpathians; viz. that the nummulitic group of Southern Europe, and which extends over such an enormous area in Asia, is the true Eocene tertiary of geologists. The cretaceous or uppermost secondary rocks of the north of Spain consist of two great stages, the lower of which is the Diceras limestone, and the uppermost a group of limestones and argillaceous sandstones, &c. with Hippurites, Radiolites and *Orbitolites*. The last-mentioned bodies have been supposed to be Nummulites; and hence has arisen the mistake of supposing, that Nummulites and Hippurites are associated in those limestones of the south which represent the chalk of the north of Europe. Above the zone of Orbitolites, is a yellowish limestone with Spatangi, which representing the upper chalk of the north, is widely developed at Santander, between the town and the lighthouse.

The nummulite limestone then follows as the next deposit in ascending order, and is overlaid, as in the Alps, by sandstone, &c. In this formation M. de Verneuil discovered, in addition to Nummulites, the *Serpula spirulæa, Conoclypus conoideus, Ostrea crassissima* or *gigantea*, fossils so well known in the nummulite rocks of the Alps, Vicentine and Crimæa. This eocene group, whose fossils are so distinct from those of the cretaceous system, nevertheless follows all the flexures and dislocations of the latter, just in the manner recently described by Sir Roderick Murchison in the Alps and Apennines. The same relations, zoological and stratigraphical, are said (on the authority of Don Amalio Maestre the Inspector of Mines of the Asturias) to extend from Aragon towards Valencia.

In describing the principal features of the carboniferous rocks of the Asturias (some of the peaks of whose limestones rise to upwards of 8000 feet above the sea), M. de Verneuil shows, that the chief seams of coal are fairly intercalated with courses of limestone and schists charged with the well-known British species *Productus antiquatus, P. punctatus* and various marine fossils. In this and in other overlying stages with conglomerates, &c. containing coal, there is, the author observes, no sandstone or schist which can have served as a soil on which jungle or marsh plants can have grown; and seeing the alternation of the fossil vegetables with marine deposits, he concludes that these coal-fields, like many others, and particularly those of the Donetz in Russia described by Sir R. Murchison and himself, were formed in estuaries of the sea by the transport and subaqueous deposit of terrestrial spoils, and are not referable to the same origin as certain carboniferous strata of the British Isles, America, &c., the coal beds of which are supposed to have been formed of vegetable masses *in situ*. In the second stage of this carboniferous formation, M. de Verneuil discovered, that courses of calcareous schists were loaded with Fusulinæ—a point of very great interest; since these foraminifera have been described in the mountain limestone of Southern Russia*, and were subsequently discovered by M. de Verneuil in the carboniferous limestone of the United States of America. Their occurrence at this intermediate station in Spain is therefore highly interesting in extending our acquaintance with the uniformity of distribution of animal life in the palæozoic ages. The coal-fields of the Asturias (of which there are seventy workable seams) seem therefore to be subordinate to the mountain limestone, like those of the north of Northumberland, the south of Scotland, &c. &c.

The Devonian system has been found to abound in the north of Spain, chiefly through the researches of M. Paillette, who has transmitted many of its fossils to France, where they have been described by M. de Verneuil.

The Triassic and Jurassic systems are also stated to be considerably developed in Spain, and like the palæozoic rocks they are highly dislocated.

In conclusion, the author remarks, that the interesting region of the Asturias will soon be better known, first through a very exact geographical map prepared by M. Paillette, particularly in reference to its coal-fields; and next by a general geological map of the province by Don. G. Schultz, on which that gentleman has been occupied during four years, and which is spoken of as a work of great merit.

* See Russia in Europe and the Ural Mountains, vol. i.

Printed in the United States
By Bookmasters